ENVIRONMENT AND HOUSING IN THIRD WORLD CITIES

ENVIRONMENT AND HOUSING IN THIRD WORLD CITIES

Edited by
Hamish Main *and* Stephen Wyn Williams

JOHN WILEY & SONS
Chichester · New York · Brisbane · Toronto · Singapore

Published by John Wiley & Sons Ltd,
Baffins Lane, Chichester,
West Sussex PO19 1UD, England
Telephone National Chichester (0243) 779777
International +44 243 779777

Other Wiley Editorial Offices

John Wiley & Sons, Inc., 605 Third Avenue,
New York, NY 10158-0012, USA

Jacaranda Wiley Ltd, 33 Park Road, Milton,
Queensland 4064, Australia

John Wiley & Sons (Canada) Ltd, 22 Worcester Road,
Rexdale, Ontario M9W 1L1, Canada

John Wiley & Sons (SEA) Pte Ltd, 37 Jalan Pemimpin #05-04,
Block B, Union Industrial Building, Singapore 2057

Library of Congress Cataloging-in-Publication Data

A CIP catalog record for this book is available from the Library of Congress

British Library Cataloguing in Publication Data

A catalogue record for this book is available from the British Library

ISBN 0-471-94831-4

Typeset in 10/12pt Palatino by Acorn Bookwork, Salisbury
Printed and bound in Great Britain by Biddles Ltd, Guildford, Surrey

Contents

List of figures

List of tables

Preface and acknowledgements

In comparison with global issues such as acid rain, global warming and depletion of the ozone layer, the environmental problems of urban areas (and Third World cities in particular) have only recently become a major item on the environmental agenda. Fortunately, during the last few years the causes of, and solutions to, such problems have become the focus for a number of research and policy initiatives. In the latter context, for example, the environmental problems facing Third World cities were given prominence during the Earth Summit held in Rio de Janeiro in June 1992. Similarly, the World Bank is currently developing a strategy framework for urban environmental management in developing countries.

In part, the focus of this volume on environment and housing can be seen as a contribution to the relatively recent body of literature concerned with the so-called 'Brown Agenda'. In another sense, however, the chapters reflect a concern for the complex interactions between people and their environments which geographers and others involved in development issues have been researching for a number of years.

The editors wish to acknowledge the financial support of the Third World Science, Technology and Development Forum for the workshop at Staffordshire University in September 1991 where several of the contributions to this book were presented; also Staffordshire University for its financial and material support of the workshop. Frank O'Reilly of the Third World Science, Technology and Development Forum deserves credit for help in organising the workshop and for suggesting its subject. We thank Kerry Summerfield and Beth Martin of Staffordshire University's Division of Geography for their skilful interpretation and retyping of some of the contributions and Owen Tucker of the Division's Cartography Section for ably preparing and drawing several of the figures. We are grateful to Derek Pratts, who is always willing to share his understanding of environmental matters, for his helpful comments on the opening and closing chapters; to Iain Stevenson, Publisher at John Wiley & Sons, who has shown enthusiasm for the book's publication and patience during its gestation; and to Sabina Sayeed for painstakingly compiling the indexes.

Biographies

Dr Hazel Barrett lectures in geography at the School of Natural and Environmental Sciences, Coventry University. Her research interests are in agricultural production and marketing, regional inequality, urban systems and population change, and her fieldwork experience has been mainly in north and west Africa and south Asia.

Dr Cecilia Chan lectures in the Department of Social Work and Social Administration at the University of Hong Kong and is a research fellow of both the Centre of Urban Planning and Environmental Management and the Population Institute of the East–West Centre. She has worked extensively with low-income communities and grassroots organisations in Hong Kong and has been a community worker in a squatter area for seven years.

Dr Martin Degg lectures in the Department of Geography at Chester College of Higher Education. His research interests focus on earthquake hazards and development issues. Together with Professor J.C. Doornkamp, he has produced a series of Earthquake Hazard Atlases for the Reinsurance Offices Association.

Dr Howard Fox, who lectures in Physical Geography at the University of Derby, is Co-director of the University's Centre for Land Evaluation and Management. His research activities on soil erosion, hydro-biology and land-use change are strongly focused on Morocco.

Dr Virginia Jiménez-Díaz lectures in the School of Geography at the Universidad Central de Venezuela, Caracas. Her specialist interest is in applied geomorphology, and she has researched on soil erosion, oil spills and physical aspects of urban expansion in Venezuela. She also plays a practical role as technical assessor for a squatter settlement organisation in Caracas.

Malcolm Lupton is an MA graduate of the Department of Geography and Environmental Studies at the University of the Witwatersrand,

Johannesburg where he is now doing doctorate research. His research interests are the political economy of South African urbanism and the gold mining industry of the Witwatersrand.

Dr Hamish Main lectures in the Geography Division of Staffordshire University, having lectured during 1981–8 in Bayero University, Kano, Nigeria and previously at the University of Birmingham. His research interests are in social and environmental aspects of urbanism and urbanisation.

Professor Harikesh Misra is Head of the Department of Geography at Himachal Pradesh University, Shimla. A Senior Fulbright Fellow at the University of Pittsburgh in 1976–7, and Commonwealth Academic Staff Fellow at the University of Cambridge in 1988, Professor Misra has published four books and fifty research papers in Indian and international journals.

Michael Mutter, a chartered town planner and chartered architect, is head of Lewis and Hickey Town Planning, part of a broad physical environment consultancy which includes the work of the Max Lock Group, an informal liaison of planning specialists in the United Kingdom and overseas. He is now doing research work for a PhD at the University of Westminster, and is senior architectural and planning adviser for Asia and the Pacific within the Overseas Development Adminstration until 1996.

Professor Margaret Peil, of the Centre of West African Studies at the University of Birmingham where she was until recently Director, is a sociologist with particular research interests in people in the cities of Anglophone West Africa. The author of several books, she formerly lectured at the University of Ghana, Lagos University and Fourah Bay College in Sierra Leone.

Dr Deborah Potts lectures in the Geography Department of the School of Oriental and African Studies of the University of London where she has worked since 1979. She specialises in the social, economic and political geography of southern Africa. Her field research on rural–urban migration, urbanisation and urban housing has been mainly conducted in Zimbabwe, Malawi and recently in South Africa.

Dr Leslie Sklair lectures in the Department of Sociology at the London School of Economics. He has researched the role of transnational corporations in developing countries, particularly in Mexico and China, and is the author of several books on this and related subjects.

Dr Linda Stanier, who lectures in Geography at the University of Derby, is Course Leader of the University's Earth and Life Studies programme.

Her specialist interests include urban geography and medical geography, and her main research focus is on service provision for older persons.

Dr Gordon Walker, who lectures in the Geography Division of Staffordshire University, specialises in environmental and planning matters. He is currently the Secretary of the Planning and Environment Study Group of the Institute of British Geographers.

Dr Stephen Wyn Williams lectures in the Geography Division of Staffordshire University, having previously lectured at Keele University. He specialises in war and peace studies, and Third World regional development, and is now developing a research interest in China.

Tony Wolfson is an MSc graduate of the Department of Town and Regional Planning of the University of the Witwatersrand, Johannesburg. He is now a development planner for PLANACT, a Johannesburg-based professional planning consultancy for grassroots civic associations and trade unions. His research interests focus on the Witwatersrand's urban land-use patterns.

1 Introduction

Hamish Main

The central position of Third World cities within the wider global environmental concerns of the early 1990s was demonstrated at the Earth Summit of June 1992 in Rio de Janeiro, where, according to the United Nations Development Programme (1993, Foreword), there was 'broad-based agreement that the developing world's growing urban populations need attention, and that their main concern is the "brown agenda" – involving pollution problems, environmental hazards, and poverty'. European interest in Third World urban environmental issues was confirmed by the large number of newspaper articles and television documentaries appearing during the weeks leading up to the Rio Summit: on storms and floods, landslides and fires, air pollution and collapsing buildings in Third World cities, though afterwards this media spotlight soon swung away elsewhere. But a report produced a few months later, in 1993, outlined the thinking of the Urban Management Programme of the joint UNDP/World Bank/UNCHS (Habitat) coordinated effort on urban management and the environment. Recommending the development of environmental strategies and action plans for individual cities, this report and the fact that it was considered necessary by such influential organisations have gone some way towards ensuring that the 'brown agenda' will not quickly be discarded by the managers of cities around the Third World.

The 'brown agenda' for dirty, unsafe cities contrasts neatly with the 'green agenda' of natural resource use and management beyond the urban boundaries, but First World perceptions of the responses needed to Third World problems are not necessarily the same as Third World perceptions. Rio was a disappointment to many in the Third World who felt that although the summit was to be about environment and development, too much discussion was centred on the former at the expense of the latter. Fears have been expressed about a new eco-imperialism: making the South pay – in lost future development – for the North's pollution during its past development (*The Ecologist*, 1992, p. 122). Similarly, there can be little doubt that poverty and its alleviation are high on the list of concerns of most Third World urban residents, but it is at least questionable whether pollution problems and environmental hazards are greater anxieties than accommodation or health or interpersonal relationships for people in many Third World cities (see Peil, Chapter 10 in this volume).

Two issues dominate concerns for housing and the environment in Third World cities in the 1990s. One is the apparent hazardous nature of many urban environments: there has been a marked increase in rapid-onset manifestations of the natural environment like floods and landslides, which all too often prove disastrous, as well as an increase in man-made hazards and disasters such as industrial pollution and major accidents. The other issue centres on inadequate conditions of housing and infrastructure suffered by large numbers of urban residents and on the health problems which they create or exacerbate. Both of these issues have been driven, at least in part, by markedly worsening urban poverty arising from recession during the 1980s. Since the 1970s in each of the major Third World regions the number of people living in absolute poverty has increased more rapidly in urban than in rural places (Gilbert, 1992, p. 437).

The greatest environmental disasters are not cataclysmic explosions and earthquakes and floods that occasionally kill hundreds of people overnight but the infectious diseases that kill and debilitate huge numbers of people, rural and urban, around the Third World. 'Every day, the number of children that die in India is greater than the total number of people who died in the Bhopal chemical disaster' (Rice and Rasmusson, 1992, p. 71). Some infectious diseases, which have been in decline in the Third World for forty years, have shown signs of increase during the past decade or so (ibid., p. 70), not least in the cities where diseases like dysentery, typhoid, cholera and tuberculosis thrive in conditions of inadequate provision for water-supply and sanitation, overcrowding, low levels of nutrition and limited health care (Hardoy *et al.*, 1992, pp. 38–48).

It is appropriate for these overlapping issues to be presented and analysed together in this book for a number of reasons: their common environmental thread and the particular emphasis on urban residential conditions and siting; their close association in the experiences of many urban residents; the particular vulnerability of the urban poor, evident in each of these issues; their common need for more sympathetic intervention by the state and for further community organisation; and because some of the responses of individuals and communities, as well as by the state, to geo-hazards and man-made hazards are similar to the responses indicated for problems of housing conditions and infrastructure.

Changing and contrasting attitudes to the environment

In Naples, within sight of the volcano Vesuvius, it is said that 'civilisation exists by geological consent, subject to change without notice'. Consent for continuing civilisation below volcanoes may be elsewhere attributed to God, or to the gods or to nature; even modern industrial society can do little except keep out of the way. Away from such high-intensity manifestations of nature, however, there has been an arrogance in the attitude of industrial society towards the environment, an arrogance that sees the environment predominantly as a resource to be exploited for profit, without care for damage that may be caused in the process. Towards the end of the twentieth century changes in popular First World attitudes to the environment may stem mainly from perceptions of the threats posed to all of us by ozone depletion, global warming and associated potential calamities, though they may also be part of a broader post-industrial realignment in attitudes. The radical environmentalism of Deep Green groups has its wider counterpart in more modest advocacies of stewardship that view the environment in a context of responsibility to future generations (Sarre *et al.*, 1991, p. 34). But a perceptible increase during recent decades in environmental disasters, in their incidence and in the damage they cause, has also contributed to 'a popular belief that we are moving into an age of disaster, and this may, to a degree, be the truth' (McCall, 1992, p. 1).

The changes now occurring in the attitudes of Third World people to the environment differ in some important ways from those of First World people. It would be an oversimplification to explain these changes as driven by the needs of industrialising societies in general, along the lines of First World societies in a previous era, in contrast to current deindustrialisation and associated attitudinal changes in the latter. But the need to exploit their resources for the purposes of development, in a manner that is not too heavily constrained by First World concerns for the global environment, is widely felt by individuals and their governments. As Dr Mahathir Mohamad, Prime Minister of Malaysia and a leading voice on these issues, has said, peasants are likely to be far less concerned about holes in the ozone layer than about the holes in their roofs (*Panorama*, 1992). Greater concern for the environment arises in the Third World from perceptions of more immediate environmental problems.

Two major developments in the analysis of environmental hazards arose out of events during the 1970s (Smith, 1992, pp. 4–5). First, particular attention was paid to the Third World, which had been suffering far more than its fair share of environmental disasters (Harrison, 1981, pp. 29–31) and to the relationship between underdevelopment and environmental hazards. Second, a series of industrial disasters in the First World during the mid-to-late 1970s (Flixborough, UK; Seveso, Italy; Three Mile Island, USA) focused attention on man-made environmental hazards. The two perspectives merged after a horrifying spate of industrial disasters in Third World cities in 1984 (in Cubatão in Brazil, Mexico City and Bhopal in India), which killed more people than had all previous technological disasters since 1945 together (on the latter two see Walker, Chapter 3 in this volume).

Most of the damage wrought by these industrial disasters was inflicted on nearby residential neighbourhoods, and this raised several questions. Why were houses built so close to hazardous industrial installations? Why had precautionary municipal planning not averted the threats to these neighbourhoods? Why were Third World societies, urban as well as rural, apparently becoming so much more vulnerable? Why was this happening in the 1980s in particular?

Third World urban environments

Urbanisation and the environment

Urbanisation is a two-edged sword: it is a means by which rural people can escape undermined rural economies or damaged rural environments, but is itself a major contributor to this environmental damage. Rising urban demand for food and declining numbers of farmers catalyse agricultural intensification that, through reducing fallow and clearing tree-cover, all too often creates and exacerbates environmental damage. Urban demand for water and energy, and for space to dump urban wastes, puts additional pressures on rural environments (Stren and White, 1989, pp. 7–9; Satterthwaite, 1991, p. 122).

All of this is relevant to an understanding of the environmental implications of Third World urbanisation and is pertinent context for some of the issues discussed in the following chapters, but the environmental arenas indicated in the preceding pages are of more immediate concern for this volume. Natural hazards, and the environmental disasters in which they may culminate, are predominantly rural problems, but rapid urbanisation and related aspects of Third World development have increasingly made them features of concern for Third World cities towards the end of the twentieth century. The lack of necessary physical infrastructure and services – water-supply, sewerage, drainage, refuse disposal and so on – become especially critical with the rapid growth of large cities.

Environmental processes, environmental hazards and environmental disasters

Up to 100,000 earthquakes may be detected annually around the world by seismometers, but without such sensitive instruments humans are aware of perhaps 1,000 of them; one hundred might cause damage costs, and ten might be capable of catastrophe, but only one or two actually cause a major disaster (Jones, 1991, pp. 29–30). Thus manifold environmental processes yield comparatively few hazards and rare disasters.

Environmental hazards may be either rapid-onset, intensive events such as earthquakes and landslides and industrial explosions or slow-onset, pervasive processes like rising groundwater and the spread of atmospheric pollution. Some environmental hazards originate in natural environmental processes – volcanoes, hurricanes, etc. – which would occur regardless of local human activity, while others like the effects of industrial production and transport pollution are clearly man-made (McCall, 1992). But many environmental hazards, with a mixed natural and man-made genesis, are natural processes which have been accentuated by human activity (Jones, 1991, p. 35; 1993, p. 162): floods in Lagos and Bangkok, and landslides in Caracas and Rio de Janeiro, have been stimulated by the manner of residential or other urban expansion, even if they have been triggered by weather or other natural events. As environmental hazards in Third World cities are more closely examined, the distinction between natural and man-made causes becomes increasingly blurred.

Some idea of the vast and complex realm of environmental hazards may be gained from Figure 1.1. This is merely a starting-point, however, which could be extensively elaborated upon. Floods, for example, might result directly from dam construction or other human activity; from landslides or changes in groundwater arising from human activity (from global warming in the most extreme case); from weather storms, seismic or geomorphological events. Which hazards are classified as 'mixed' and which are classified as 'natural' or 'man-made' is in part open to interpretation (Jones, 1993, p. 162).

A clear need exists, then, to identify potential links between environmental processes, environmental hazards and environmental disasters; especially to establish how human activity may encourage or discourage the transformation of a process into a disaster. It might be argued that humans could avoid many environmental hazards by staying away from regions of high-intensity natural processes such as seismic belts along the edges of tectonic plates and zones of tropical windstorm tracks. It is, after all, only if people are involved that environmental processes become environmental disasters. Yet major concentrations of urban and rural settlement, in parts of the First World as well as the Third World, follow the continental fracture zone around the Pacific Rim and the lands threatened by tropical windstorms in southern Asia and the Caribbean. The environmental hazard is often well understood, but it is outweighed by the perceived benefits of continued residence: highly fertile volcanic earths and floodplain alluvia have been created by the same eruptions and storms and floods that threatened previous residents (Rapp *et al.*, 1991, p. 214).

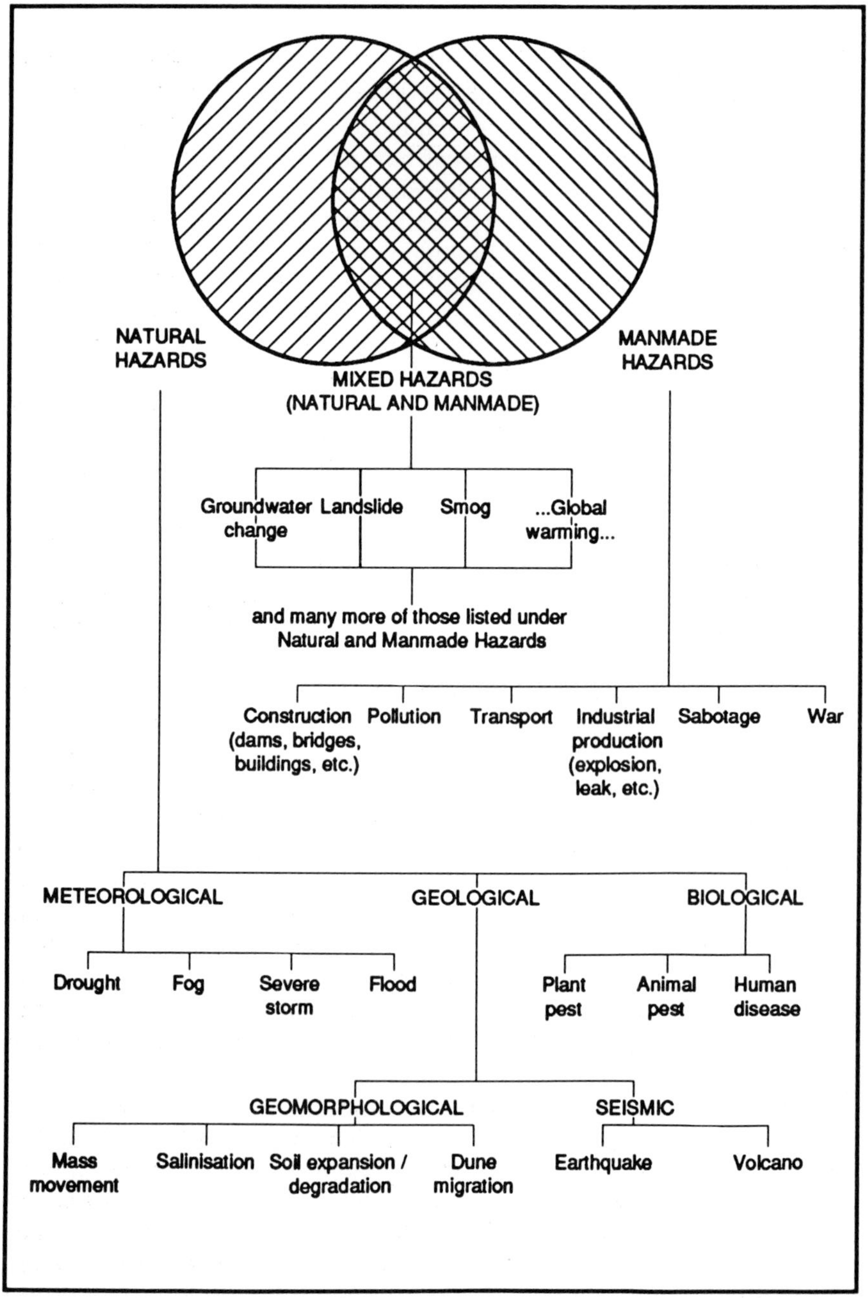

Figure 1.1 Environmental hazards in Third World cities: a classification. Derived from Jones, 1991, p. 35: McCall, 1992, pp. 1–3; Jones, 1993, p. 162.

Urban environmental disasters and the enhanced vulnerability of the urban poor

Three seismic disasters that struck within two months in late 1992 killed over 1,800 people, nearly all of them in urban places. The epicentre of the Egyptian earthquake in October 1992 was twenty miles south of Cairo: Degg (Chapter 2 in this volume) shows how the large number of casualties in the capital (over 550 dead and 10,000 injured) was in part a product of unsafe building practices. The Egyptian government quickly announced measures to curb such building practices, already illegal but widely used. But it seems unlikely that government measures like these will be effective, as long as urban poverty creates demand for cheap housing well beyond the government's ability to provide it. When a volcano erupted from Cacaqual Mountain in north-western Colombia, six days after the Egyptian earthquake, the nearby town of San Pedro de Uraba was showered with boiling mud and rock: thirty people were reported to have died in San Pedro and at thermal springs outside the town where they were bathing. Over 1,230 lives were lost in December of the same year when an earthquake epicentred in the sea off Indonesia's Flores Island triggered a tidal wave (tsunami). Most of the casualties were in Maumere, a town of about 70,000 people, where 80 per cent of the buildings were destroyed including most of the residential areas.

A clear tendency exists today for urban residents to be far more vulnerable than nearby rural people to the initial impact of extreme weather and seismic events and to the landslides and floods and building collapses set off by them. Some at least of the people who live in cities are more constrained to hazardous sites, and dense settlement and multistorey buildings without the necessary shock-proof structure multiply this vulnerability. Of over a hundred people killed in the August 1993 tropical storm that swept over Venezuela from the Caribbean Sea, more than three-quarters were in Caracas, where dwellings were demolished by landslides in several of the squatter neighbourhoods indicated by Jiménez-Díaz in Chapter 7. This point is also well made by Degg (Chapter 2 in this volume) with respect to the Mexico City earthquake of 1985.

Whereas a relatively small number of major seismic disasters accounts over the years for large numbers of casualties in Third World cities, and these often receive extensive coverage in the international media, a much larger number of extreme weather events goes without widespread coverage. Those events indicated in Table 1.1 are probably by no means representative of such recent events in Third World cities. An apparent spate of urban disasters induced by extreme weather during June–July 1992, according to Table 1.1, might reflect climatic conditions at that time of year in some Third World regions, but their inclusion may well have

Table 1.1 Some urban environmental disasters arising from severe and extreme weather events in the Third World, February 1992 – August 1993

Date	*Weather event*	*City*	*Disaster*
February 1992	severe rain and snow	East Jerusalem	cemetery wall, undermined by rain and snow, fell on roof of cafe, killing twenty-four and injuring twenty
April 1992	heavy rain	Kigali, Rwanda	landslide left 200 people homeless and disrupted water supply; no deaths or injuries reported
May 1992	torrential rain	Hong Kong	landslides and floods killed five people; thousands of homes evacuated
June 1992	torrential rain	Dushanbe, Tajikistan	mudslides destroyed houses, killing up to ten people
June 1992	torrential rain	Antofagasta, Chile	mudslide destroyed large area of low-income housing; at least sixty-four dead, 750 injured and 20,000 homeless
June 1992	freak rainstorms	Chibuto, Mozambique	mudslides drowned or buried over fifty people while the town was under attack by right-wing terrorists
July 1992	tornado	Panama City	dozens of buildings in the city's outskirts collapsed, killing at least seven and injuring fifteen people
July 1992	heavy rain and wind	Hong Kong	buildings damaged, one person killed and eleven injured
July 1992	heavy rain	Chittagong, Bangladesh	landslide buried homes, killing six people
September 1992	flashfloods	Manila, Philippines	mudslides killed at least four people
April 1993	violent storm	Massawa, Eritrea	2,000 'shacks' destroyed, three people killed and 10,000 homeless
April 1993	torrential rain	Port Sudan, Sudan	rain and flood destroyed 'shanties and houses', and killed at least twenty-one people

Table 1.1 *(continued)*

Date	*Weather event*	*City*	*Disaster*
May 1993	hurricane, flashfloods, waterspouts	Dushanbe, Tajikistan	widespread damage; 'waves of mud' destroyed about 3,000 homes; thousands homeless
August 1993	tropical storm	Caracas, Venezuela	landslides killed over 100 people, injured hundreds and made thousands homeless in squatter neighbourhoods

Sources: Ceefax (BBC television) and Oracle (ITV), various dates

more to do with the attention of the media on the Rio Environment Summit of early June and heightened awareness at that time of environmental matters in general.

Such disasters are often measured in small numbers of deaths, if any, but may create extensive homelessness and damage to property. Cities in many Third World regions suffer their effects (see Table 1.1). The construction of low-income neighbourhoods in hazardous sites, and the flimsy nature of housing in such neighbourhoods, clearly contributed to several of these disasters. Some of the features of the Antofagasta disaster are common to cities elsewhere in the Third World. Until about twenty-five years ago, when Antofagasta had developed little beyond its colonial core, vegetation on the hillsides above the town would have reduced the impact of occasional torrential rains. Subsequent construction of extensive low-income neighbourhoods has expanded the built-up area up the hillslopes, most of the vegetation has been consumed as fuel and large water storage tanks have been constructed to supply urban demand (Vidal, 1992, p. 27). A torrential rainstorm on 16 June 1992 triggered a huge mudslide that toppled the water storage tanks, thereby adding to its own momentum, and ripped through low-income neighbourhoods. 'In short, Nature did what it has always done: man brought this one, literally, on himself' (Vidal, 1992, p. 27).

The Antofagasta disaster illustrates well the point that distinctions between 'natural' and 'man-made' disasters are usually blurred. Environmental disasters in and around industrial production zones, oil pipelines, sewerage systems, rubbish dumps, borrow pits, burial grounds, airport flight-paths, electricity cables and other residential infrastructure are more obviously man-made and show a strong urban bias because of the predominantly urban nature of land uses like these. The spate of industrial disasters in 1984 drew attention to the fact that unsafe industrial practices no longer tolerated in the First World were threatening Third World neighbourhoods (see Walker, Chapter 3; and Sklair, Chapter 5, in this volume).

Table 1.2 Some man-made urban environmental disasters in the Third World, December 1991 – August 1993

Date	*City*	*Disaster*
December 1991	La Grita, Venezuela	high-voltage power cable collapsed on Christmas street party and killed six people
February 1992	Kwangju, South Korea	three gas tanks exploded; nine people were injured and over 1,000 families evacuated
March 1992	Dhaka, Bangladesh	fire in a low-income neighbourhood killed five people and made 10,000 homeless
April 1992	Guadalajara, Mexico	industrial waste products dumped in sewers caused explosions that devastated large areas of the city, killed over 170 people and required evacuation of over 70,000
April 1992	Delhi, India	explosions in the old city destroyed three buildings and killed thirteen people
June 1992	Casablanca, Morocco	eight people were crushed to death and seven others injured when their old three-storey house collapsed
July 1992	Tbilisi, Georgia	overloaded cargo plane failed to gain sufficient height on takeoff and hit block of flats in residential area, killing over twenty-seven people in flats as well as all thirteen on board
April 1993	Istanbul, Turkey	methane gas explosion at the city's biggest rubbish dump buried a dozen shanty dwellings, killing at least sixteen people and burying up to one hundred
August 1993	Nakhon Ratchasima, Thailand	hotel building collapsed, killing 128 people, after four extra floors and water storage tanks for 150,000 litres had been added to the original two-storey building

Sources: Ceefax (BBC television) & Oracle (ITV), various dates

These 1984 industrial disasters all wrought terrible destruction on nearby low-income neighbourhoods, and this is a feature common to industrial environmental disasters as higher-income neighbourhoods tend to be located away from industrial zones and associated utilities. This kind of locational bias is also evident to some extent in most other types of man-made environmental disaster in Third World cities (see Main and Williams, Chapter 9 in this volume). Table 1.2 gives recent examples of such tragedies: it is probably fair to assert that the great majority of urban environmental disasters caused by fire, explosions without known cause or building collapse due to unspecified structural failure go unreported by the international media.

The vulnerability of the urban poor to environmental hazards has been emphasised by much of the analysis of the past decade, since Susman *et al.* (1983) offered a marginalisation theory of disaster that pointed to the nature of the society in which disaster occurs and in particular the differential vulnerability of social classes: 'Without people, there can be no disaster. And poor people are generally more vulnerable than rich ones. Disaster is therefore defined as the interface between an extreme physical event and a vulnerable human population' (p. 264). This structural or political economy analysis has proved more convincing for many authors than the behavioural form of analysis which dominated much of the previous literature on environmental hazards, explaining disasters as arising from ineffective social adjustment to the physical environment and advocating largely technological responses to hazards (Jones, 1993, p. 163).

Structural approaches to explanation of environmental hazard have their critics, however. 'The blanket assertion that the poor are most at risk from natural calamities assumes that there are a large number of rich, rational people about, with perfect scientific knowledge at their fingertips, who are able to push their way past their poorer compatriots to avoid the risky places' (Pearce, 1991, p. 21). It was after the 1976 earthquake in Guatemala, which killed over 22,000 people, that the term 'classquake' was used to refer to its patently differential effects in the capital city, where damage was 'almost exclusively in the slum areas' (Susman *et al.*, 1983, p. 275). Pearce (ibid.) argues that 'the risk here is that a single case will be converted into an immutable law by virtue of a neat turn of phrase'. But the evidence does support the view that the poor suffer disproportionately from environmental disasters. As in housing and other arenas, the strength of a structural approach to environmental hazards lies in its broad-brush explanation of why the urban poor tend to be most damaged by hazard events. It is less potent in indicating the detail of how the events occur or of what, given the nature of the societies in which they occur, should be done to minimise the damage.

Third World urban housing

Third World housing issues were dominated until about the 1970s by a preoccupation with securing an adequate supply of housing units of sufficient quality. When it became clear that subsidised public-sector housing in city after city, region after region was failing to achieve such housing goals, attention turned to a redefinition of what constituted sufficient quality. Inappropriate colonial housing standards were replaced by those more in keeping with the requirements and abilities of the urban masses, and this encouraged various means of housing provision including prefabrication, core houses, sites-and-services schemes and housing cooperatives (Devas, 1993, p. 80). Some real progress was achieved in a number of cities, but, just as public housing programmes had done, these alternatives tended to benefit only the better-off among the urban residents. So the central problems remained and indeed worsened as urbanisation proceeded apace: urban population growth exceeded the accommodation constructed, and large numbers of the urban poor in particular continued to be excluded from access to legal housing. Thus squatter settlements and other informal and/or illegal accommodation came to shelter many of the urban poor, typically between one-third and two-thirds of the population of cities in many Third World regions (Hardoy and Satterthwaite, 1989, pp. 25 and 76–7). Even in a city like Agadir, with considerable state investment in and controls on housing from the time of the earthquake in 1960 up to 1985, *bidonvilles* still housed 18 per cent of the population in 1985 (Barrett *et al.*, Chapter 4 in this volume).

Squatters came to realise that their neighbourhoods were less likely to be demolished or threatened by the authorities if they were sited in areas that were undesirable for legal private or public development. They increasingly occupied the hillslopes and floodplain margins and the peripheries of the industrial estates and airports (see Main and Williams, Chapter 9 in this volume). 'Poorer groups may even choose particular locations where pollution levels are highest because these are the only locations where they can find land for their housing close to sources of employment without fear of eviction' (Hardoy *et al.*, 1992, p. 107). The residents of Chheetpur, the Allahabad squatter settlement described by Misra (Chapter 11 in this volume) had been evicted twice from previous neighbourhoods, so moved to Chheetpur specifically because it was a swampy site.

The pioneering work of Mangin (1967), Turner (1969) and others in Latin America, where many of these problems were encountered before other Third World regions, offered the real possibility of a way around the impasse of squatter settlements: they suggested that, by legalising instead of demolishing squatter settlements, and by supporting their residents' efforts to improve their housing structures and neighbourhood infrastructure, municipal authorities could turn the apparent *problem* of squatter settlements into a *solution* to the real problem of insufficient urban housing supply – which, after all, would merely be aggravated by demolition. A spate of squatter upgrading programmes in the late 1970s and early 1980s, many of them supported by international donor agencies, put these ideas into practice in cities such as Lusaka and Jakarta, though elsewhere such as in Santiago and Mexico City (Gilbert, 1991, p. 289) and Lagos major squatter demolitions have continued through the 1980s.

That municipal authorities have not taken to improving their illegal and other informal settlements more wholeheartedly than they have done in the past twenty years may be explained by several reasons. It is more costly to upgrade neighbourhoods that have grown piecemeal than to provide the infrastructure (water pipes, drains and sewerage, roads etc.) prior to house construction, and Third World municipal budgets are as a rule very limited. Legalising the illegal occupation by squatters of land claimed as property by landowners is unlikely, given the usual shortcomings of compensation, to prove popular with a social group that tends to have close contacts with municipal authorities. On the other hand, if the landowners are able to take rents and other fees from the squatters, this may be an attractive option for some landowners so long as the squatters' tenure is not legalised – and especially if the squatter settlement is demolished subsequently when the time is right for sale or development, thereby yielding the profits of land speculation in addition to those of rent. Another cornerstone of early capitalist development, the availability of local cheap labour, also encourages the continued existence of squatter settlements rather than their demolition or legalisation. Thus squatter settlements may serve various functions for capitalist development in their ambivalent situation: illegal and threatened with demolition but not yet legalised or demolished (Peattie, 1979, p. 1021; Main, 1990, pp. 15–17).

Although much has been achieved in squatter upgrading programmes and in reducing the threat of demolition, one of the effects of this trend in urban housing has been to concretise the demotion of the urban poor to neighbourhoods in environmentally or locationally marginal sites. Municipal authorities that operate on behalf of the interests of landowners and industrialists are most unlikely to upgrade and legalise squatter settlements sited on urban land that is attractive for development.

Considerable variation has been evident within the Third World in the efforts of the urban poor to gain access to adequate accommodation and in state responses to these efforts. More repressive municipal attitudes to squatters, and more regulated controls over rural–urban migration and the means of access to land and housing, have been typical of the former settler colonies in southern Africa, for example, than elsewhere in the continent (Amis, 1990, pp. 15–16). This is clear in a comparison of the descriptions in this volume of urban housing in southern Africa (Lupton and Wolfson, Chapter 6; Potts, Chapter 12) and in West Africa (Peil, Chapter 10; Mutter, Chapter 13).

The most evident trend during the 1980s in Third World urban housing has been the withdrawal of the state from housing provision, though the effects of this have varied. It has meant a pragmatic tolerance by the state towards informal neighbourhoods in some societies (on Hong Kong, see

Chan, Chapter 8 in this volume) where, as self-help housing has expanded, so owner-occupation has developed rapidly for the urban poor. But higher rates of owner-occupancy are limited eventually by shortages of available land, as pools of peri-urban public and communal land dry up. Thus in some cases informal land access is now becoming less widespread: there are fewer organised land invasions around Latin American cities, while communal land around many African cities has practically all been developed (Gilbert, 1992, p. 440). Land shortages may also continue to result from strict government controls on land access or from topography and other natural features: steep slopes (on Caracas see Jiménez-Díaz, Chapter 7 in this volume), floodable land and so on. In some such cities, rental accommodation has expanded rapidly in recent years.

Clearly there are advantages and disadvantages for urban residents, poor as well as better-off, in the state's reduced role in housing provision. As for the provision of urban infrastructure, the disadvantages appear to outweigh the advantages (Stren, 1990, p. 50; Gilbert, 1992, pp. 444–51). Although few Third World states as yet have formally implemented such a policy, it has been occurring in effect for twenty years or more – as the needs of low-income neighbourhoods in particular have been ignored by municipal authorities, so residents and informal enterprises have not only built houses but also supplied water, disposed of waste and provided unofficial link-ups to mains electricity (Stren, 1988, p. 243; Gilbert, 1992, p. 449).

Issues of housing and environment in Third World cities

For those who can afford to do so, the choice to reside away from the industrial zone or the steep slopes or the riverbank might be made in order to avoid the industrial pollution or the daily crowds or the problems of access over difficult terrain, but probably this is taken for granted. They may or may not be aware of the hazardous industrial processes within the factories or the threats of landslides or floods. Rarely would they consider the possibility of a future environmental disaster there. Overriding concerns with new accommodation might be less to do with the neighbourhood's site than with the quality of its housing and facilities and perhaps even most to do with its social esteem. But in ensuring that their new house is sufficiently comfortable and well supplied, and their new neighbourhood sufficiently prestigious to maintain their social standing (this will be clear from the price), most if not all of them will also ensure – without realising it – that, should the unforeseen disaster strike, they will suffer the consequences far less than those who cannot afford to live in their neighbourhood and must make do with a shack behind the factory or on the floodplain.

Increasingly in Third World cities the invisible hand of the market deals out environmental safety along with comfort and prestige, environmental hazard along with discomfort and disregard. There are exceptions: the Mexico City earthquake disaster of 1985 wrought most tragedy among the lower-middle class in high-rise buildings (Degg, Chapter 2 in this volume), and high-income Lagos residents suffer the same flooding as low-income Lagosians (Peil, Chapter 10 in this volume). But, as a rule, there is no need for better-off urban residents to have 'perfect scientific knowledge of risk at their fingertips' in order 'to avoid the risky places' (Pearce, 1991, p. 21 – see above): it is the urban poor who suffer most from environmental disasters and hazards, as they do from the ongoing lack of adequate housing and infrastructural facilities. Chan's chapter on Hong Kong is one of several in this volume that demonstrate this. Lupton and Wolfson show in Chapter 6 how the apartheid South African state has directed the relegation of large portions of the Witwatersrand urban poor to a range of hazardous residential environments along racial lines, and pose the question how a similar process will be achieved after apartheid by the market.

Inadequate urban infrastructure

Just as Third World governments and municipal authorities have increasingly lacked the resources to deal with the sheer shortage of housing, so they can rarely afford to install the water-supply pipes and drains and sewerage systems that, with growth in city populations and rising residential densities, are increasingly needed. It is ironic that rising awareness of the need for more adequate urban infrastructure, which has been heightened by attention to Third World environmental matters in the past decade, should have occurred at the same time as municipal powers have been reduced and public housing programmes have ground to a halt in many Third World cities and the private sector has been given a freer hand in the supply and maintenance of many urban services (Amis, 1990, pp. 20–4; Devas, 1993, pp. 98–9). Scope exists for this increasing reliance on market forces to do great damage to the residential conditions, health and livelihoods of people in cities around the Third World – and it will inevitably be those who are least able to protect their own interests, the urban poor, who suffer most.

A lack of adequate urban infrastructure means more than just inconvenience for those in neighbourhoods without piped water, drainage and sewerage or reliable refuse disposal. Urban growth and industrialisation carry with them the risk that unpiped water supplies will become polluted by informal sewage disposal or by the careless disposal of solid or toxic wastes. Excessive urban extraction of groundwater is causing buildings in

Mexico City and Bangkok to sink, while it has allowed salt water to penetrate groundwater sources and thereby foul drinking-water supplies in a number of cities including Lima, Dakar, Tel Aviv and Manila (Meadows *et al.*, 1992, p. 57). Mexico City is only the best known of dozens of Third World cities with dangerous levels of atmospheric pollution, usually caused by emissions from vehicles or industries or domestic fires. Chinese and Indian cities are prominent among many Third World cities with levels of suspended particulate matter several times that recommended as a maximum by the World Health Organisation (World Resources Institute, 1988, p. 169).

Poor housing quality is the most visible sign of a low-income neighbourhood, especially where the threat of eviction reduces the incentive to invest in physical improvements, but more hardship may be caused for the neighbourhood's residents by the lack of infrastructure and services. Exceedingly rapid population growth in Nouakchott during the early 1980s was partly caused by especially harsh environmental conditions in the rural interior of Mauritania. By the mid-1980s, two-thirds of the city's residents were without direct access to water and had to buy their supplies from water merchants: this water was of variable quality and cost nearly a hundred times as much as that used by people with piped connections to the mains (Harpham *et al.*, 1988, p. 35). The case of Nouakchott in the 1980s was exceptional, but the principle holds true for many of the urban poor around the Third World: not only do they have to make their own arrangements for domestic water, sewage, electricity and so on, but they are also liable to end up paying more than people in better-off neighbourhoods for these infrastructural services. The subsequent informal arrangements for sewage- and rubbish-disposal, electricity link-ups and so on inevitably make such residents more vulnerable to infectious diseases and accidents. And this vulnerability is enhanced by their neighbourhoods' sites, adjacent to factories or polluted water courses and additional poverty-related problems like unemployment (ibid., pp. 41–2).

Some success has been achieved with ventilated improved pit latrines as cheap, hygienic means of sewage disposal in cities such as Harare and Gaborone (Potts, Chapter 12 in this volume). But where residential population densities are high, or neighbourhoods' environmental conditions are awkward (rocky ground conditions, perhaps, or a high water table), there may be no viable hygienic alternative to relatively expensive sewerage networks.

Historical antecedents

Respect for potentially damaging environmental processes has been evident in urban management for thousands of years. The Minoans of

Crete incorporated a variety of anti-earthquake construction techniques into their cities over 3,500 years ago. It has been postulated that the huge volcano of about 1628 BC that blew half of the nearby island of Santorini into the sea also brought the Minoan civilisation to an end, either directly through tsunami and/or perhaps earthquake or indirectly through the devastation of agriculture by volcanic ashfall (Renfrew, 1979). But the direct effects of this volcano were apparently survived by Knossos, the Minoan capital. Although located on the north coast of Crete, facing Santorini, Knossos was built far enough inland to be out of reach of the tsunami's most damaging effects. Some of its houses were built of hewn stone, while others were constructed three or four storeys high on the sides of hills, rather than on valley floors which would be more vulnerable to seismic shock as well as floods. Wooden beams and joints were incorporated into house construction to increase elasticity against earthquakes.

The Incas of the Andes also built their settlements on hillsides, away from valley floors vulnerable to earthquake and lahar (that is volcanic mudslide). Their towns were deliberately restricted in size, with buildings sited well apart from each other. Low walls, thatched roofs and an avoidance of low beams positioned directly overhead reduced the vulnerability of those inside crashing buildings (Pearce, 1991, p. 21).

Like the colonial Spanish, who obliterated pre-Columbian civilisations and ignored the accumulated wisdom of their urban management techniques, other European colonisers around the globe also tended to assume there was nothing to be learnt from the architecture and urban forms of colonised societies (though there were exceptions, as in India). By the time of mass industrialisation, environmental constraints to urban growth in Europe were ignored where possible or otherwise minimised.

The appalling conditions of water supply, environmental pollution, sanitation and overcrowding in Britain's industrial towns were revealed by housing and sanitary inquiries during the 1840s and narrated graphically by social commentators like Engels, whose vivid descriptions of a dozen industrial towns around the British Isles at this time show how widespread urban environmental degradation was, how closely it was associated with industrial development and how damaging it was to health (Engels, 1982, pp. 60–106).

It is hardly surprising that urban mortality was far higher in industrial cities than in other urban and rural areas or that cholera epidemics like those of the 1840s devastated the populations of such districts. The life expectancy of Manchester's working-class people was only seventeen years at that time, compared with thirty-eight years in rural England; and, though Manchester's gentry and professional people had more than double (thirty-eight years) the life expectancy of the working class, this was also appreciably lower than that of their rural counterparts' fifty-two years (Knox, 1987, p. 18).

It is striking that commentators like Engels were concerned with environmental degradation less because of its aesthetic desolation than for the health hazards it engendered among those who suffered it most. Governments in Britain and elsewhere in industrialising Europe, representing the perceived interests of the industrial bourgeoisie, resisted calls for reforms to reduce these environmental health hazards. But the clear association between these degraded residential conditions and health problems, both endemic and epidemic, encouraged the formation of public associations pressurising government to take appropriate steps to improve public health: legislation for and implementation of hygiene regulations, paved streets, housing standards, and publicly funded water and sewerage systems. The Health of Towns Association, founded in 1844 and soon with branches in thirteen British towns, disseminated data from official reports, organised public lectures and meetings and petitioned Parliament. The Association's effectiveness as a pressure group can be gauged by the 1848 Public Health Act, which at last began to respond through legislation to existing urban environmental conditions (Ashton, 1992, p. 2) and by subsequent national legislation and municipal action (Douglas, 1989, p. 227).

Health and safety initiatives in Third World cities today

Clearly, there are too many differences in scale and technology for ancient Minoan and Inca cities to be a model for Third World municipal management today. The intention in presenting this material here is to point to what can be done with basic materials at the scale of the individual dwelling or neighbourhood. Recent community-based hazard mitigation programmes have been drawn up in Latin American cities like Lima, but government support for such initiatives is lacking (Maskrey, 1989, pp. 41–6).

Although major differences obviously exist between the present situation in some Third World cities and that in early industrial British cities, parallels are evident too – especially the need for an environment-conscious approach to urban health care. The environmental focus of urban health care demonstrated by the 1848 Act remained central to health policy until late in the nineteenth century, when the germ theory of disease moved the emphasis towards personal prevention through immunisation. This was superseded in turn, from the 1930s, by a therapeutic era when treatment with pharmaceutical 'magic bullets' by professional medics in large hospitals was emphasised. This latter orientation has also been the public-health model imposed on Third World societies since about the middle of the twentieth century. Yet the public-health 'problems that were described in British cities 150 years ago are all to be found in [Third World] cities today, albeit on a much bigger scale and with much greater consequences'

(Ashton, 1992, p. 6). The clear need for a much greater environmental focus in Third World urban public-health policy has been recognised by the World Health Organisation in a series of initiatives that have incorporated environmental as well as personal preventative and therapeutic elements. The most notable of these is the Healthy Cities Project, which is already bearing fruit through community organisation for environmental health improvements in a number of cities in Latin America and southern Asia (Rice and Rasmusson, 1992).

Urban managers and policy-makers might eventually be encouraged to take action on urban environmental problems at least partly out of self-interest, as in Britain's nineteenth-century cities, given the effects of unsanitary urban conditions on their health as well as on that of the urban poor (Amis, 1990, p. 14). Sklair (Chapter 5 in this volume) points to the environmental degradation of *maquila* production in Mexico's border cities being presented in the American media as a threat to the health and welfare of people north of the border.

Even where they are high among government priorities, however, environmental sanitation programmes may be misused and bring more detriment than benefit. Epidemics of bubonic plague in South Africa, first in Cape Town in 1901 and then in other large cities, were used as a pretext for the eviction of non-white urban residents to remote sites, some of them on levelled refuse tips or adjoining sewage disposal works. Thus began municipal apartheid: non-whites, labelled a threat to the health of whites, forced into unsanitary environments where many of them suffered appalling health problems and escalating mortality rates (Swanson, 1977; Beavon, 1982). In contemporary as well as colonial Nigeria, as Stock (1988) has shown, the urban poor were 'blamed for living in unsanitary environments and threatening public health' (p. 30) while inevitably the elites benefited most from environmental health programmes. The Nigerian state's Environmental Sanitation campaign of the mid-1980s continued to ignore industry's sordid role in environmental pollution while the clearance of squatter accommodation, roadside traders and other informal activity, and various other projects having little or nothing to do with the protection of the people's health were legitimated by the 'sanitation syndrome'.

The challenge in the Third World now, according to Ashton (1992, p. 7), 'is to move towards cities and towns which are based on sound ecological principles', that minimise intrusion into the natural environment by taking account of topographic, hydrological, vegetational and climatic conditions. This is a daunting challenge, but it advocates precisely the same principles as those practised by the Minoans, the Incas and other ancient civilisations. A more recent parallel may be drawn with the town planning principles of Patrick Geddes and the Garden City movement of the early twentieth century in Britain which were also a reaction to the environmental degradation of early industrialisation and mass urbanisation. Yet there seems little prospect of effective environment-conscious urban planning on this sort of scale in the Third World, especially when, as in the 1990s, the state is withdrawing from urban management.

Literature review

Few books specifically on Third World urban environmental processes or problems were published before the late 1980s. Some regional geographies included material on the environments of individual cities, such as Hance (1970) and O'Connor (1983) on African cities, but systematic treatment of urban environmental processes was sporadic. Detwyler and Marcus (1972), a notable exception, explained neglect of urban people's interaction with their physical environments as resulting from the prevailing Western view of the city as almost entirely a cultural phenomenon: 'As cities have grown, so has man's local domination over nature and with it the idea that man is independent of nature . . . just as natural scientists have tended to "assume away" man, so too have social scientists tended to "assume away" environment' (Marcus and Detwyler, 1972, p. 9). It is instructive that just four of the 54 chapters in the six volumes on *Geography and the urban environment: progress in research and applications*, edited by Herbert and Johnston and published between 1978 and 1984, focused on features of the natural environment and only one of these referred to Third World material.

Ian Douglas's *The Urban Environment* (1983) heralded the new era of urban environmental concerns. Including an array of material on cities around the Third World, it examined associations between urban geomorphological and biological processes and human activity in a detailed manner rarely seen outside the occasional journal paper. John Doornkamp's sustained interest in the geomorphology of cities in the Middle East and elsewhere, and its application to construction and planning processes, has been reflected in a series of publications during the 1980s, among them Cooke *et al.* (1982) and Doornkamp (1985).

The environmental hazard literature had yet to focus consistently on cities, but themes that later became important in the Third World urban literature were already evident in the 1970s. In explaining 'natural hazards' as hazards only when people were involved, and in examining whether greater environmental damage and larger environmental catastrophes were likely to result from the nature of contemporary development, Burton *et al.* (1978) helped to set the agenda for much of what followed. Sheets and Grayson (1979) again sought to encourage the study of natural hazards and disasters by social scientists, who had done little in this field, as well as by natural scientists who had done much more.

A tremendous quantity of empirical data on disasters existed, but theoretical analysis by social scientists remained weak until the early 1980s. The volume edited by Hewitt (1983) contained several useful theoretical interpretations, including those of Watts and of Susman *et al.*, who pointed to the vulnerability of the poor; this has been a continuing focus for much of the work published recently. Maskrey's 1989 book was exceptional in examining disaster mitigation measures by community organisations which he advocated as more effective than government responses and likely to lead to more permanent reductions in the vulnerability of the poor. A systematic and detailed treatment of risk, hazard and disaster in the environment was offered by Smith (1992).

The journal *Environment and Urbanization* began publication in 1989 with issues on environmental problems (1,1) and housing for the poor (1,2) in Third World cities and has maintained an ongoing focus on these subjects. Other journals, like *Regional Development Dialogue* (1991), have devoted special issues to Third World urban environmental items. Jorge Hardoy and David Satterthwaite, from the editorial board of the former journal, have been in the forefront of recent work on urban housing and environmental matters. Among a large number of publications resulting from their collaboration, a suite of four books will perhaps be the most influential. The first of these (Hardoy and Satterthwaite, 1989) pointed to the difficulties of access to housing faced by many of the Third World's urban poor and outlined in two seminal chapters the associated problems of urban environments. This environmental concern was elaborated (Hardoy *et al.*, 1992) in an extended version of a report produced by the authors for the Rio Summit; the new environmental agenda for Third World cities outlined in the book presaged the UNDP's 'brown agenda'. The book edited by Hardoy *et al.* in 1990 had demonstrated close associations between health problems and environmental and housing conditions. Each of these three books scrutinised the role of government, usually ineffective, in dealing with the urban problems identified and suggested alternative means by which these problems might be tackled at the community level. The fourth book, not yet published, is to develop the latter theme: 'the need for new forms and channels of funding and technical support from governments and international agencies towards community level initiatives' (Hardoy *et al.*, 1992, p. 12).

Housing, infrastructure and health have been identified by Hardoy and Satterthwaite in these books as the three most salient concerns associated with environment for Third World urban residents. Housing, recently reviewed by Gilbert (1992), has been the chief interest of the Third World urban literature during the past thirty years. Dwyer (1975) and Gilbert and Ward (1985) were exceptional prior to the late 1980s in pointing to urban environmental problems. Land was a major focus in the 1980s (Angel *et al.*, 1983; Payne, 1984; McAuslan, 1985). Problems and management of urban infrastructure and servicing have recently been represented in Stren and

White (1989), Amis and Lloyd (1990) and Devas and Rakodi (1993). The health of the urban poor, an issue neglected until the mid-1980s (but see Agarwal *et al.*, 1981), has been the concern of a number of recent publications that make a clear linkage with environmental conditions (Harpham *et al.*, 1988; Ashton, 1992).

Structure of the book

The book's organising framework has been provided by the major issues regarding Third World urban environments and housing. Environmental disasters, both natural and man-made, are analysed with reference to residential conditions in the first Section, and environmental hazards are analysed similarly in the second. The third Section focuses on ongoing environmental concerns of urban housing and infrastructure. The inevitable overlap between these issues is reflected in the middle Section, where urban hazard is examined in the sense both of the potential for environmental disaster and of existing problems for residents and neighbourhood development.

State involvement in issues of environment and housing is a common concern that runs through all three Sections, though differences in authors' emphasis are apparent. The need for strong physical planning controls, to reduce the threats of environmental hazards to urban residents, is argued in each of the opening Section's chapters; to argue otherwise would hardly be tenable where major disasters have occurred recently and might well happen again. This need is echoed in some of the subsequent chapters, but in places it is set more in a context of why such planning controls are not effectively implemented by some Third World cities' managers, or (as in the Republic of South Africa) why they are predominantly implemented to achieve ends very different from protection against environmental hazard.

The final contributed chapter presents the environment as resource, a positive feature for urban development; it is fitting that it has been written by Michael Mutter, whose vision as an architect has not been blunted by the cynicism of the geographers and sociologists responsible for the portrayal of environment in a largely negative light through most of the book. Williams's Conclusion seeks appropriate and realistic responses to the problems illustrated in the preceding chapters, given the constraints facing Third World states and people in the 1990s.

References

Agarwal, A., Kimondo, J., Moreno, G. and Tinker, J., 1981, *Water, sanitation, health – for all? Prospects for the International Drinking Water Supply and Sanitation Decade, 1981–90*, London, Earthscan.

Amis, P., 1990, 'Key themes in contemporary African urbanisation' in P. Amis and P. Lloyd (eds), *Housing Africa's urban poor*, Manchester, Manchester University Press, pp. 1–34.

Amis, P. and Lloyd, P. (eds), 1990, *Housing Africa's urban poor*, Manchester, Manchester University Press.

Angel, S., Archer, R., Tanphiphat, S. and Wegelin, E. (eds), 1983, *Land for housing the poor*, Singapore, Select Books.

Ashton, J. (ed.), 1992, *Healthy cities*, Milton Keynes, Open University Press.

Beavon, K.S.O., 1982, 'Black townships in South Africa: *terra incognita* for urban geographers', *South African Geographical Journal*, 64, pp. 3–20.

Burton, I., Kates, R.W. and White, G.F., 1978, *The environment as hazard*, New York, Oxford University Press.

Cooke, R.U., Brunsden, D., Doornkamp, J.C. and Jones, D.K.C., 1982, *Urban geomorphology in drylands*, Oxford, Oxford University Press.

Detwyler, T.R. and Marcus, M.G., 1972, *Urbanization and environment: the physical geography of the city*, Belmont, Duxbury.

Devas, N., 1993, 'Evolving approaches' in N. Devas and C. Rakodi (eds), *Managing fast growing cities: new approaches to urban planning and management in the developing world*, Harlow, Longman, pp. 63–101.

Devas, N. and Rakodi, C. (eds), 1993, *Managing fast growing cities: new approaches to urban planning and management in the developing world*, Harlow, Longman.

Doornkamp, J.C., 1985, *The earth sciences and planning in the Third World*, Liverpool Planning Manual, no. 2, Liverpool, Liverpool University Press.

Douglas, I., 1983, *The urban environment*, London, Arnold.

—, 1989, 'The rain on the roof: a geography of the urban environment' in D. Gregory and R. Walford (eds), *Horizons in human geography*, Basingstoke, Macmillan.

Dwyer, D.J., 1975, *People and housing in Third World cities*, London, Longman.

Ecologist, The, 1992, 'The earth summit débâcle', editorial in special issue: Whose common future? 22 (4), p. 122.

Engels, F., 1982, *The condition of the working class in England*, London, Granada. Originally published in German, 1845, Leipzig.

Environment and Urbanization

Gilbert, A., 1991, 'Urban problems in the Third World' in P. Sarre (ed.), *Environment, population and development*, Milton Keynes, Open University Press.

—, 1992, 'Third World cities: housing, infrastructure and servicing', *Urban Studies*, **29** (3/4), pp. 435–60.

Gilbert, A. and Ward, P., 1985, *Housing, the state and the poor: policy and practice in three Latin American cities*, Cambridge, Cambridge University Press.

Hance, W.A., 1970, *Population, migration and urbanization in Africa*, New York, Columbia University Press.

Hardoy, J.E. and Satterthwaite, D., 1989, *Squatter citizen: life in the urban Third World*, London, Earthscan.

Hardoy, J.E., Cairncross, S. and Satterthwaite, D. (eds), 1990, *The poor die young: housing and health in Third World cities*, London, Earthscan.

Hardoy, J.E., Mitlin, D. and Satterthwaite, D., 1992, *Environmental problems in Third World cities*, London, Earthscan.

Harpham, T., Lusty, T. and Vaughan, P. (eds), 1988, *In the shadow of the city: community health and the urban poor*, Oxford, Oxford University Press.

Harrison, P., 1981, *Inside the Third World: the anatomy of poverty* (second edition), Harmondsworth, Penguin.

Herbert, D.T. and Johnston, R.J. (eds), 1978–84, *Geography and the urban environment: progress in research and applications*, Volumes 1–6. Chichester, Wiley.
Hewitt, K. (ed.), 1983, *Interpretations of calamity: from the viewpoint of human ecology*, London, Allen & Unwin.
Jones, D.K.C., 1991, 'Environmental hazards' in R.J. Bennett and R.C. Estall (eds), *Global change and challenge*, London, Routledge, pp. 27–56.
—, 1993, 'Environmental hazards in the 1990s: problems, paradigms and prospects', *Geography*, **78** (2), pp. 161–5.
Knox, P., 1987, *Urban social geography: an introduction* (second edition), Harlow, Longman.
Main, H.A.C., 1990, 'Housing problems and squatting solutions in Metropolitan Kano' in R.B. Potter and A.T. Salau (eds), *Cities and development in the Third World*, London, Mansell, pp. 12–31.
Mangin, W., 1967, 'Latin American squatter settlements: a problem and a solution', *Latin American Research Review*, **2** (3), pp. 65–98.
Marcus, M.G. and Detwyler, T.R., 1972, 'Urbanization and environment in perspective' in T.R. Detwyler and M.G. Marcus, *Urbanization and environment: the physical geography of the city*, Belmont, Duxbury, pp. 3–26.
Maskrey, A., 1989, *Disaster mitigation: a community based approach*, Development Guidelines Series, no. 3., Oxford, Oxfam.
McAuslan, P., 1985, *Urban land and shelter for the poor*, London, Earthscan.
McCall, G.J.H., 1992, 'Natural and man-made hazards: their increasing importance in the end-20th century world' in G.J.H. McCall, D.J.C. Laming and S.C. Scott (eds), *Geohazards: natural and manmade*, London, Chapman & Hall, pp. 1–4.
McCall, G.J.H., Laming, D.J.C. and Scott, S.C. (eds), 1992, *Geohazards: natural and manmade*, London, Chapman & Hall.
Meadows, D.H., Meadows, D.L. and Randers, J., 1992, *Beyond the limits: global collapse or a sustainable future*, London, Earthscan.
O'Connor, A., 1983, *The African city*, London, Hutchinson.
Panorama, 1992, 'The South strikes back', BBC1 Television, 1 June 1992.
Payne, G.K., 1984, *Low-income housing in the developing world: the role of sites and services and settlement upgrading*, Chichester, Wiley.
Pearce, F., 1991, 'Acts of God, acts of man?', *New Scientist*, 18 May, pp. 20–1.
Peattie, L., 1979, 'Housing policies in developing countries: two puzzles', *World Development*, **7**, pp. 1017–22.
Rapp, A., Li, J. and Nyberg, R., 1991, 'Mudflow disasters in mountainous areas', *Ambio*, 20 (6), pp. 210–18.
Regional Development Dialogue
Renfrew, C., 1979, 'The eruption of Thera and Minoan Crete' in P.D. Sheets and D.K. Grayson (eds), *Volcanic activity and human ecology*, London, Academic Press., pp. 565–86
Rice, M. and Rasmusson, E., 1992, 'Healthy cities in developing countries', in J. Ashton (ed.), *Healthy cities*, Milton Keynes, Open University Press, pp. 70–86.
Sarre, P. and Smith, P. with Morris, E., 1991, *One world for one earth: saving the environment*, London, Earthscan, in association with The Open University.
Satterthwaite, D., 1991, 'Urban and industrial environmental policy and management' in J.T. Winpenny (ed.), *Development research: the environmental challenge*, London, Overseas Development Institute, pp. 119–38.
Sheets, P.D. and Grayson, D.K. (eds), 1979, *Volcanic activity and human ecology*, New York, Academic Press.
Smith, K., 1992, *Environmental hazards: assessing risk and reducing disaster*, London, Routledge.

Stock, R.F., 1988, 'Environmental sanitation in Nigeria: colonial and contemporary', *Review of African Political Economy*, 42, pp. 19–31.

Stren, R., 1988, 'Urban services in Africa: public management or privatisation?' in P. Cook and C. Kirkpatrick (eds), *Privatisation in less developed countries*, Brighton, Harvester Wheatsheaf, pp. 217–47.

—, 1990, 'Urban housing in Africa: the changing role of government policy' in P. Amis and P. Lloyd (eds), *Housing Africa's urban poor*, Manchester, Manchester University Press, pp. 35–72.

Stren, R.E. and White, R.R. (eds), 1989, *African cities in crisis: managing rapid urban growth*, Boulder, Westview Press.

Susman, P., O'Keefe, P. and Wisner, B., 1983, 'Global disasters, a radical interpretation' in K. Hewitt (ed.), *Interpretations of calamity: from the viewpoint of human ecology*, London, Allen & Unwin, pp. 263–83.

Swanson, M.W., 1977, 'The sanitation syndrome: bubonic plague and urban native policy in the Cape Colony, 1900–1909', *Journal of African History*, 18, pp. 387–410.

Turner, J.F.C., 1969, 'Uncontrolled urban settlement: problems and policies' in G. Breese (ed.), *The city in newly developing countries*, Englewood Cliffs, Prentice-Hall, pp. 507–34.

United Nations Development Programme/World Bank/United Nations Centre for Human Settlements (Habitat), 1993, *Towards environmental strategies for cities: policy considerations for urban environmental management in developing countries*, Strategy framework paper, New York, United Nations.

Vidal, J., 1992, 'The unnatural face of modern disaster', *The Guardian*, 21 June.

Watts, M., 1983, 'On the poverty of theory: natural hazards research in context' in K. Hewitt (ed.), *Interpretations of calamity: from the viewpoint of human ecology*, London, Allen & Unwin, pp. 231–62.

World Resources Institute/International Institute for Environment and Development/United Nations Environment Programme, 1988, *World Resources 1988–89*, New York, Basic Books.

Section 1

Disastrous environments, vulnerability and response

2 Perspectives on urban vulnerability to earthquake hazard in the Third World

Martin Degg

In the 1960s and 1970s the 'dominant view' of natural hazards was that of environmental determinism: the belief that natural disasters can be explained simply in terms of the operation of extreme, but natural physical processes. Linked with this was the assumption that the only meaningful way to do anything about such phenomena was through the application of technology (and capital) to 'fix' the environment, the so-called 'technological fix'. This view is amply illustrated in a number of important publications of the era, including those of White (1974) and Burton *et al.* (1978). The 1980s witnessed some change of emphasis in hazard and disaster studies, with Hewitt (1983) and his co-authors challenging the dominant viewpoint and arguing that socio-economic and political factors are of paramount importance in understanding why people are vulnerable to extremes in the environment and therefore experience disasters. This 'alternative view' has increasingly been incorporated into many aspects of hazard and disaster study, although social scientists have been quick to point out the disappointing lack of reference to societal processes and human vulnerability in the declared objectives of the United Nations' International Decade for Natural Disaster Reduction (IDNDR) (Mitchell, 1988). Alexander (1991, p. 213) concludes that 'in the eyes of some practitioners, the IDNDR has assumed the status of a "technofix", in which the proponents of technology and hard science use it as a justification for generating yet more of the same'.

While environmentally deterministic approaches to hazard and disaster study have received due criticism in recent years, it is important that a balanced viewpoint should prevail. In particular, social scientists should take care to ensure that the weaknesses of the environmentally deterministic approach do not become replicated in their own work through studies that are essentially 'socially deterministic' in nature. Hewitt (1983) recognised that total abandonment of the 'dominant view' would result in an equally unbalanced view of natural hazards and disasters and do little to further the understanding of the complex interplay of issues involved. A holistic, interdisciplinary approach – in which appropriate consideration is paid both to the vagaries of the physical environment as well as to the social, economic and political factors that influence human behaviour – is the only meaningful way forward for hazard and disaster research.

Set against this background, this chapter first of all considers the global distribution of earthquake hazard in order to set the environmental context. It then examines some of the factors that influence the vulnerability of the Third World city, as a unit, to the earthquake hazard and discusses appropriate response strategies. The chapter draws heavily upon loss experience from recent and historical earthquakes in various parts of the Third World.

Exposure of the largest urban areas to earthquake hazard

As far as environment and housing in Third World cities is concerned, the earthquake hazard is particularly significant because of its association with increasingly catastrophic losses in recent decades and because of the important role played by housing in influencing the vulnerability of people to the hazard – the actual physical process of ground shaking poses little threat to human beings sitting out-of-doors in the open. Figure 2.1 shows zones of earthquake hazard and the related hazards of volcanic and tsunami ('tidal' wave) activity.

Variations in hazard intensity within the zones are not shown on the figure owing to limitations of scale but are described elsewhere (Degg, 1992a). The hazard exposure data have been derived from the *World Map of Natural Hazards* (scale 1:30,000,000) produced by the Munich Reinsurance Company (1988). Onto the figure have been plotted the location of the world's one hundred largest cities, as defined by the United Nations (1989). These cities range in size from the giant metropolises of the Western countries (for example Tokyo/Yokohama and New York with populations of 19 million and 15.6 million respectively) and the primate cities of the Third World (for example Mexico City and São Paulo with populations of 16.6 million and 15.5 million respectively) to the city of Pune (India) with a population of 1.1 million. In 1985 these cities accommodated 10 per cent of the world's total population and were divided between the Western and Third World in the ratio 2:3.

An analysis of the data in this figure shows that 60 per cent of the largest cities in Western countries are exposed to one or more of the hazards on the map but that 78 per cent of the largest Third World cities are in this position. On an overall hazard exposure scale ranging from zero (no exposure to any of the hazards in the figure) to seven (maximum

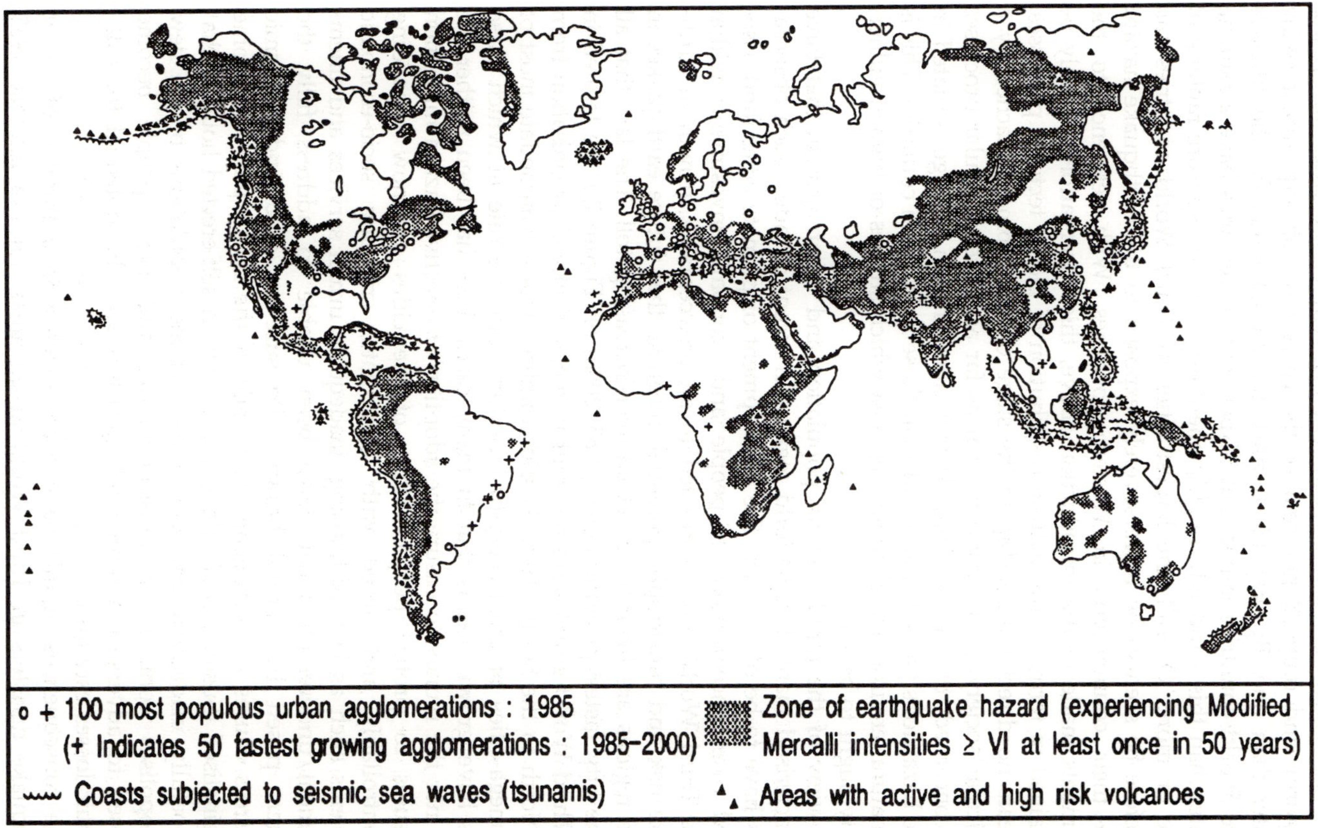

Figure 2.1 The environmental context: major zones of earthquake activity relative to the world's 100 largest urban areas (Source: amended from Degg, 1992a)

exposure to all of the hazards), the modal hazard exposure value for the cities of Western countries is zero (40 per cent of the cities are in this category), and only 38 per cent of the cities have an exposure ranking greater than one. By way of contrast, the modal value for the cities of Third World countries is one, but 54 per cent of the cities have a ranking greater than this – indeed, roughly one-third have an exposure ranking of three or more. Overall, the largest cities in Third World countries can therefore be seen to be more severely exposed to the earthquake hazard than their counterparts in Western countries. With reference to the comments made in the introduction above, this observation can only be explained in terms of the natural distribution of global tectonic processes. It bears no direct relationship to the stage of development reached by a particular society; it just happens that as far as these particular processes are concerned, the largest Third World cities are, on average, situated in more hazardous parts of the world than those in Western countries. This is the natural environmental context against which studies of human vulnerability ought to be set.

In view of the rapid rates of population and urban growth that have characterised many Third World regions in recent decades, such observations have to be of concern. The development of primate cities (and associated over-concentrations of people and economic investment) within some Third World regions has led to an unacceptable growth in earthquake risk and catastrophe potential, meaning that future earthquakes in these regions are likely to exact an ever-increasing toll. The fact that all of the fifty fastest-growing cities (1985–2000) on Figure 2.1 are located in the Third World emphasises the shift in focus of global urbanisation from the North to the South. Third World countries will accommodate most of the mega-cities of the twenty-first century, but the rate at which these cities have grown means that in many cases there has been neither the time nor the money to ensure the adoption of effective hazard mitigation measures to protect people. Many of the cities are now expanding uncontrollably and, faced with immediate, crippling socio-economic problems such as lack of housing, inadequate urban services and unemployment/underemployment, have been unable to address fully the dangers posed by natural hazards. Pressure on land and economic resources causes an increasing percentage of the urban poor to become marginalised (Susman *et al.*, 1983) and to live in self-erected housing that is woefully inadequate and vulnerable. As the populations of the new metropolises continue to burgeon, more and more people become exposed to adverse environmental processes and the potential for catastrophic losses escalates.

The consequences of this have been manifest in a number of recent earthquake disasters in Third World urban areas: for example the 1985 Mexican earthquake, in which at least 5,000 people were killed and

Table 2.1 Loss experience in selected recent earthquakes, 1989–1990

Date	Location	Fatalities (homeless)	Damage (US$m)* (insured loss)*	Earthquake magnitude
21.08.88	India/Nepal	>1,052	250	6.7
06.11.88	Yunnan Province (China)	730 (500,000)	–	7.6
07.12.88	Armenia	25,000 (500,000)	14,000	6.9
23.01.89	Tadzhikistan	>284	25	6.0
01.08.89	Irian Jaya Province (Indonesia)	120	–	5.6
18.10.89	San Francisco (USA)	63	7,000 (1,000)	7.1
19.10.89	Shanxi Province (China)	29 (60,000)	–	6.1
28.12.89	Newcastle (Australia)	10	792 (396)	5.5
26.04.90	Qinghai Province (China)	>126	–	6.9
29.05.90	N.E. Peru	>200	–	6.2
21.06.90	Gilan Province (Iran)	50,000	8,000 (>115)	7.4
16.07.90	Luzon Island (Philippines)	>1,600	>370 (55)	7.7
13.12.90	E. Sicily (Italy)	>20	500	4.7

*unadjusted figures for economic losses

damage of US$ 4,000 million was caused (unadjusted figure), and the Cairo earthquake of 1992 in which more than 560 people were killed and almost 6,500 injured (Degg, 1993). The disproportionately high death tolls associated with earthquakes in Third World regions, as compared to those affecting cities in the Western world, are emphasised by Table 2.1.

Vulnerability of dwellings to earthquake hazard in Third World cities

When addressing the issue of vulnerability of housing to earthquake hazard in Third World cities, images of poor, overcrowded and inadequate shelters in squatter settlements and dilapidated houses in slums tend to predominate. It is certainly true that many of the horrific death tolls associated with earthquakes in Third World regions during the last thirty years have been attributable to such factors. For example, the Guatemalan earthquake of 1976 wreaked havoc in the slums of Guatemala City and the poor rural highlands of the country, killing 22,000 people. The phrase 'classquake' was coined by Guatemalan peasants to describe the impact of the earthquake upon the country (O'Keefe *et al.*, 1976). Similarly, but on a greatly reduced scale, the magnitude of the losses associated with the Cairo earthquake of 1992 (see above) was partly attributable to the atrocious housing conditions in which many of the poorer urban dwellers are forced to live. In view of the fact that the Cairo earthquake was of only moderate magnitude (5.9 on the Richter Scale), the human losses seem unacceptably high. Table 2.1 shows that the earthquake which affected Newcastle in south-east Australia in 1989 was of comparable magnitude and type but resulted in no more than ten fatalities. Such observations are understandably evocative, and in recent years much of the socio-political debate regarding Third World urban vulnerability to natural hazards has focused on the exposure of the poorest urban dwellers.

Although the plight of the urban poor is extremely important, it is also important not to be blinded by dogma into believing that the issue of vulnerability to natural hazards in Third World cities extends no further than the limits of the shanty towns or inner-city slum areas. As far as the earthquake hazard is concerned, when the vulnerability of *entire* Third World cities is considered, then a range of other important factors comes into play. The poor need not necessarily always be those most exposed to the hazard, and it is increasingly possible to envisage earthquake scenarios in which other sectors of the community are the worst affected. The issue is a complex one, and to take too narrow a stance is to present as unjust an analysis of Third World urban vulnerability as some of the technocratic, Eurocentric analyses produced twenty and thirty years ago. The crucial issue is that massive urbanisation is taking place on an unprecedented scale, and at unprecedented rates, in many parts of the world that are subject to more extreme environmental conditions than those found throughout much of the so-called 'developed world'. Unless meaningful measures are sought to lessen the economic, political and social burdens under which Third World cities strive to exist and grow, hundreds of thousands of lives will be lost in the coming decades. Whether these fatalities are in the self-erected houses of the squatter settlements, the inadequately designed and hurriedly constructed public buildings (for example schools, hospitals and workplaces), or the high-rise tenement and apartment blocks produced by private 'enterprise' for the middle classes, would seem to be somewhat less significant than (and almost a distraction from) the enormity of the overall problem.

Factors influencing the vulnerability of Third World cities (in their entirety) to the earthquake hazard

A number of factors are important in influencing the collective vulnerability of Third World cities to the earthquake hazard. These include:

— accessibility;
— construction type and height;
— emergency and public services.

Each of these is discussed in turn below.

The role of accessibility

In urban vulnerability studies emphasis is often placed upon the influence of standards and types of construction. Accessibility, whether at the national, regional or local level, is a factor that is seldom directly referred to, despite its important role in influencing susceptibility to loss. As far as the earthquake hazard is concerned, there are two main dimensions:

— localised accessibility within urban areas influences the ease with which people can escape shaking buildings without the risk of impedance or injury from falling masonry and the ease with which local emergency services can reach affected areas;
— regional accessibility between different parts of a given country is important because it governs the speed and effectiveness of post-disaster response. The sooner that relief can reach an affected region, the greater the chances of saving injured or trapped people and of preventing further human losses in displaced populations.

The importance of the latter has recently been emphasised by the 13 March 1992 earthquake at Erzincan (population 150,000), a provincial capital in eastern Turkey. Turkey is one of the most seismically active countries of the Mediterranean region. Seismicity in the country is concentrated along three main tectonic belts, namely the North Anatolian fault zone (which crosses the northern half of the country from east to west), the East Anatolian fault zone (predominantly in the south-east of the country) and the fault-bounded graben structures of western Turkey. Sandwiched

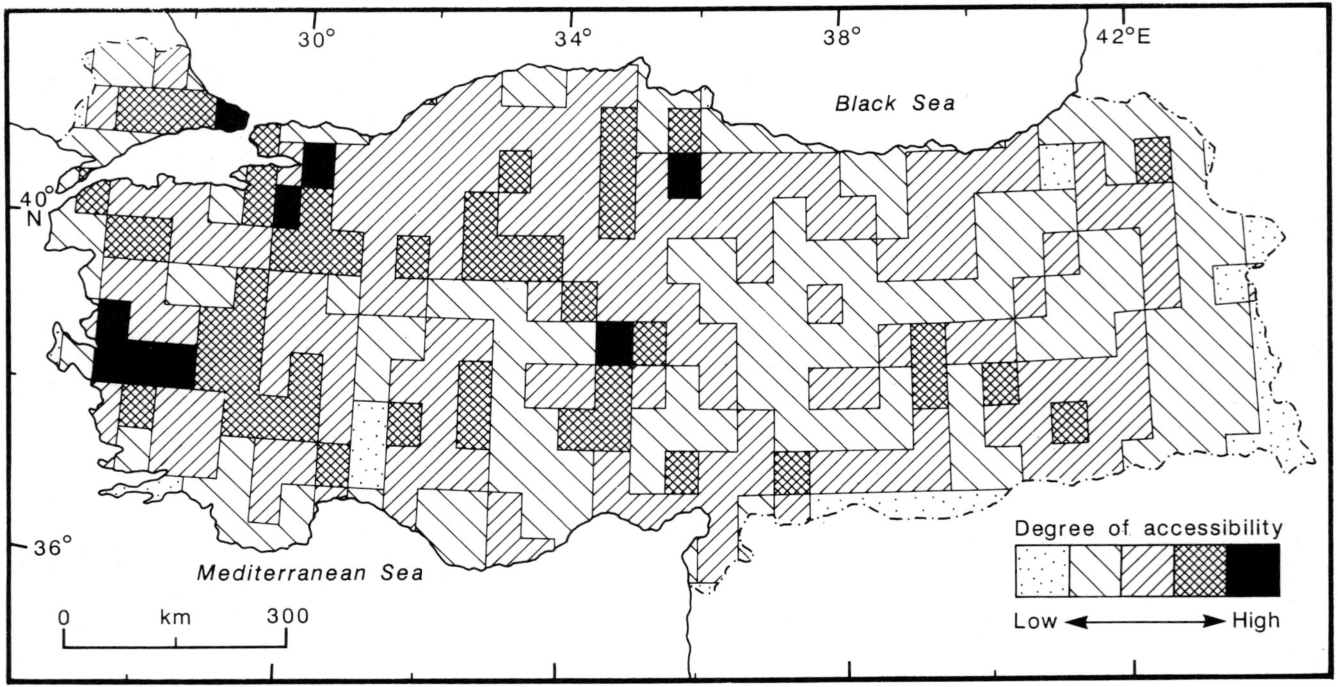

Figure 2.2 Regional accessibility by road within Turkey (Source: Kolars, 1982)

between these fault zones, central-southern Turkey is relatively stable (Degg and Doornkamp, 1992). The 1992 earthquake occurred along the North Anatolian fault, approximately 40 kilometres to the south-east of Erzincan and was of magnitude 6.8. The earthquake was felt across fourteen provinces in eastern Turkey, including in the cities of Diyarbakir, Elazig, Erzurum and Malatya, but the worst effects were confined to the province of Erzincan where the city and many remote mountain villages were badly affected. In Erzincan approximately 200 buildings experienced total collapse (tall buildings in the city centre were particularly badly affected) and 2,000 were damaged, killing up to 500 people, injuring over 600 and rendering many thousands homeless. The impact of the event was heightened by freezing weather conditions and snow-covered ground, together with the fact that a state hospital, medical school and the police headquarters collapsed, as did numerous other public buildings. Two hotels also collapsed, together with ten apartment buildings and office blocks.

The earthquake triggered many avalanches, rockfalls and landslides which blocked roads and cut the main rail link to Erzurum (and the nearest airport), 130 kilometres to the east. Relief efforts were consequently severely hampered, a factor which undoubtedly served to accentuate the severity of human losses experienced in both urban and rural areas, as well as compounding the misery of the injured and homeless. The poor accessibility of eastern Turkey, attributable to the combined effects of dispersed population centres and poor transportation infrastructure, is emphasised in Figure 2.2. This shows that for comparable hazard exposure and building type, the inhabitants of many urban and rural settlements in the eastern part of the country are more vulnerable to the effects of earthquake impact than are those in the west.

The influence of construction type and height

As discussed above, in recent years much of the socio-political debate concerning Third World urban vulnerability to natural hazards has focused on the inadequacies of the dwellings in which the poorest urban dwellers live and on the political, social and economic mechanisms which impel people to live in such conditions. It is, however, important to realise that as far as the earthquake hazard is concerned, other types of building in Third World cities may prove equally (if not more) vulnerable to certain types of ground motion and that the potential for catastrophic loss through their failure may be very much greater than that associated with the dwellings of the poorest inhabitants.

A good illustration of this point is provided by another relatively recent earthquake, that which devastated parts of Mexico City in 1985. This earthquake measured 8.1 on the Richter scale and occurred off the Pacific coastline of Mexico, over 350 kilometres to the south-west of the city. Despite this, over 5,000 in the city died as a result of the earthquake, 50,000 were injured and 250,000 were rendered homeless. The earthquake caused extremely selective destruction in the central part of Mexico City where 7,400 buildings were damaged, of which 1,665 were severely affected and 770 were total losses. Contrary to what is sometimes suggested, it was not the poorest urban dwellers in Mexico City that experienced the largest human losses from the earthquake – if anything, the lower-middle class was the single social group worst affected because the buildings that experienced the highest incidence of damage were medium- to high-rise, of between six and twenty storeys (Degg, 1989). Buildings of this elevation proved much more sensitive to the low-frequency ground motions emanating from the distant earthquake (generally speaking, high-frequency seismic energy is rapidly attenuated with increasing distance from an earthquake epicentre, leaving energy of predominantly low frequencies of vibration), motions that were amplified by the soft subsoils on which the central and eastern parts of the city stand.

The residential complex of Tlatelolco, situated just to the north of the centre, was particularly badly affected. The estate covered 150 hectares and consisted predominantly of high-rise structures; twenty-four out of the 130 buildings were rendered unsafe leaving thousands of families homeless, while part of the thirteen-storey Nuevo Leon building collapsed killing most of its one thousand inhabitants. Several other (predominantly lower-middle class) high-rise residential buildings collapsed in the Benito Juarez housing estate to the south of the centre.

It would be untrue to suggest that low-rise, poor quality housing was not affected by the Mexican earthquake: damage to such accommodation was heavy in some areas, particularly in the Morelos District to the north of the centre where old adobe and brick-built houses crumbled. However, it is probably fair to say that the magnitude of the losses associated with this type of building was nowhere near as catastrophic as that associated with the modern, tall structures. In addition, the extensive shanty towns on the outskirts of Mexico City were affected only to a minor degree; and the light, flimsy nature of the dwellings meant that where collapses did occur injuries were generally light. Pearce (1991, p. 21) concludes that 'if there was a sociological pattern in the Mexico quake, it was that public buildings – high rise flats of varying rents, hospitals and schools – suffered worst'. The impact of the failure of these public services upon people of *all* social classes should not be overlooked and is analysed in greater detail in the section below.

While the vulnerability of tall buildings in Mexico City to the 1985 earthquake represents an extreme case (owing to combined effects of the nature of the event itself and of the particular characteristics of the subsoils beneath the centre of the city), the greater vulnerability of high-rise structures to distant earthquakes has important implications for urban development and planning in the Third World. Primarily this is because the last forty years have witnessed a dramatic growth in the amount of high-rise construction taking place in Third World cities. This increased building elevation means that urban areas situated within seismic zones (see Figure 2.1) may start to experience a greater frequency of earthquake damage than that observed in the past. Others, situated on the fringes of seismic zones and beyond, may well start to 'feel' earthquakes and experience damage for the first time in their history. Hazard, risk and vulnerability assessments must, therefore, now consider the threats posed to cities many hundreds of kilometres away from the most active seismic belts.

The importance of this factor can be illustrated with reference to the eastern Mediterranean region in general and the 1992 earthquake in northern Egypt in particular. Egypt has a relatively well-documented earthquake history, with records of earthquake damage that extend back as far as 2800 BC (Kebeasy, 1990). Although not a country of high earthquake hazard, parts of Egypt are exposed to sufficient earthquake threat to warrant precautionary planning measures to reduce the risk of loss (Degg and Doornkamp, 1992). Figure 2.3 shows that the northern part of the country in particular is exposed to potentially damaging levels of hazard (see Tables 2.2 and 2.3). The Nile delta stands out as an area of higher hazard because of the combined effects of localised faulting, proximity to major tectonic zones (such as the Red Sea and its northern extension into the Gulf of Suez rift) and the presence of unconsolidated (and in some areas water-saturated) subsoils likely to provide a less secure foundation during earthquakes than consolidated sediments and solid rock. The earthquake that struck Cairo in 1992 served to emphasise the moderate levels of seismic activity that are generated locally in this part of the country.

The earthquake was of magnitude 5.9 and occurred about 20 kilometres to the south-west of Cairo. The size of the event, the area affected and the urban death toll per-capita ratio bear many similarities to an earthquake that affected the city on 7 August 1847. Ambraseys (1991) has re-evaluated this event and concludes that the epicentre probably lay to the south-west of Cairo between the city and Lake Qârûn (Faiyûm district) and that the earthquake was of magnitude 5.5–5.9. The 1847 earthquake affected predominantly old adobe and brick houses of two and three storeys, many of which were in a dilapidated state of repair. Newly built European-style houses and public buildings in the city were hardly affected at all. A total of twelve people were killed in Cairo and its suburbs.

Similarly, in 1992 poorly constructed and maintained concrete, brick and adobe two-, three- and four-storey buildings performed badly. In the sprawling slums of Cairo overcrowding forces people to live in packed,

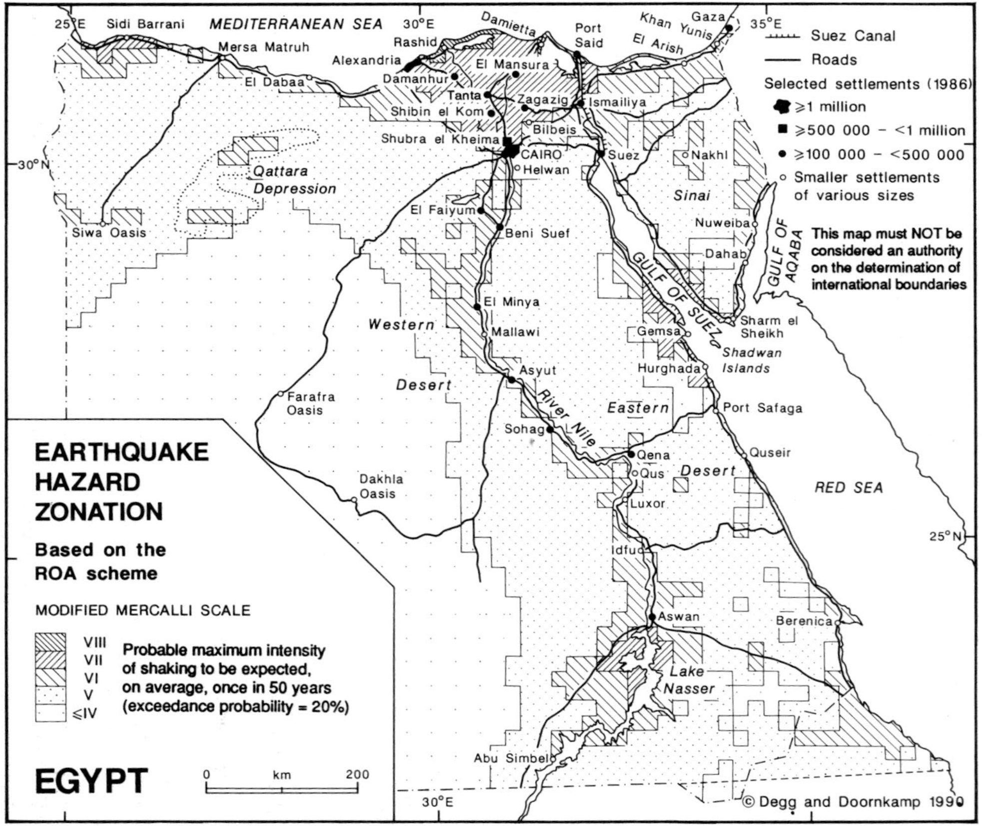

Figure 2.3 Earthquake hazard zonation of Egypt (Source: Degg and Doornkamp, 1990)

Table 2.2 The Modified Mercalli (MM) earthquake intensity scale

Modified Mercalli Scale	
I	Imperceptible
II	Felt on upper floors
III	Objects swing
IV	Felt by everybody, creaking noises
V	Felt by everybody, objects move
VI	Cracks appear in buildings
VII	Fissures in roads, difficult to stand, large bells ring
VIII	Steering on cars affected, chimneys and towers fall
IX	Landslides, underground pipes break
X	Most masonry and frame structures destroyed, serious damage to large structures, dams and dykes
XI	Serious damage to large structures and buildings
XII	Total destruction, distortion of the Earth's surface

Table 2.3 Earthquake loss-susceptibility data for different construction types

Construction type	Average damage (%) at intensity (MM) VI	VII	VIII	IX	X
1 Adobe	8	22	50	100	100
2 Unreinforced masonry, non-seismic design	3.5	14	40	80	100
3 Reinforced concrete frames, non-seismic design	2.5	11	33	70	100
4 Steel frames, non-seismic design	1.8	6	18	40	60
5 Reinforced masonry, medium quality, non-seismic design	1.5	5.5	16	38	66
6 Reinforced concrete frames, seismic design	0.9	4	13	33	58
7 Shear wall structures, seismic design	0.6	2.3	7	17	30
8 Wooden structures, seismic design	0.5	2.8	8	15	23
9 Steel frames, seismic design	0.4	2	7	20	40
10 Reinforced masonry, high quality seismic design	0.3	1.5	5	13	25

Source: Sauter and Shah (1978)

badly built tenement blocks, to which floors are often illegally added to accommodate more people. In a city where the population now approaches twelve million (compared to approximately 250,000 in 1847), there simply are not sufficient resources to provide secure housing for a large proportion of the population. The poorest of the city's inhabitants, compelled to erect makeshift shelters on sacred ground in the 'City of the Dead', were, however, largely unaffected by the earthquake. Indeed, the single greatest human loss occurred with the collapse of a fourteen-storey apartment block in the affluent northern suburb of Heliopolis. This building was originally eight storeys high but had been added to twice to give a further six storeys. It accommodated a fast-food store, a bookshop and a launderette on the ground floor and seventy flats on the floors above. Overall 120 people are estimated to have died in the building, although the death toll would have been very much higher had the earthquake occurred at night, instead of just after 3.00 p.m., local time.

Emphasis in the reporting of the 1992 Cairo earthquake by the press tended to slant towards the plight of the poor living in dilapidated three- and four-storey houses throughout the city's slums (for example Helm, 1992; Pugh, 1992). Important as this is, an equally important issue and possible cause for grave concern is the failure of the fourteen-storey concrete apartment block in what was, after all, a moderate local earthquake of shallow focal depth (less than seventy kilometres below the surface). The level of concern rises when one stops to consider that the delta region of Egypt is occasionally affected by distant earthquakes of damaging intensity. In particular, the historical earthquake record for the eastern Mediterranean region shows that high-magnitude, intermediate-depth (seventy–300 kilometres below the surface) events of the southern Aegean have been felt across parts of the delta on a number of occasions (Degg, 1990). It is the greater depth of the intermediate events that tends to result in their being felt over much larger distances than shallow earthquakes of comparable magnitude. For example, Figure 2.4 outlines the approximate area of perceptibility of the Rhodes earthquake of 26 June 1926. This earthquake occurred off the south-east coast of Rhodes, was of magnitude 7.5 and had a focal depth of approximately 100 kilometres (Papazachos and Papazachos, 1989). The areas of greatest damage were on Rhodes, Crete and in south-west Turkey, but the earthquake caused alarm and light damage around most of the eastern Mediterranean coastal margin. Damaging intensities of ground motion were experienced across much of the Nile delta, including Cairo where old buildings were affected to a minor degree (Ambraseys and Melville, 1988). The unconsolidated sediments which underlie the delta area probably served to increase sensitivity to the types of ground motion emanating from the distant earthquake.

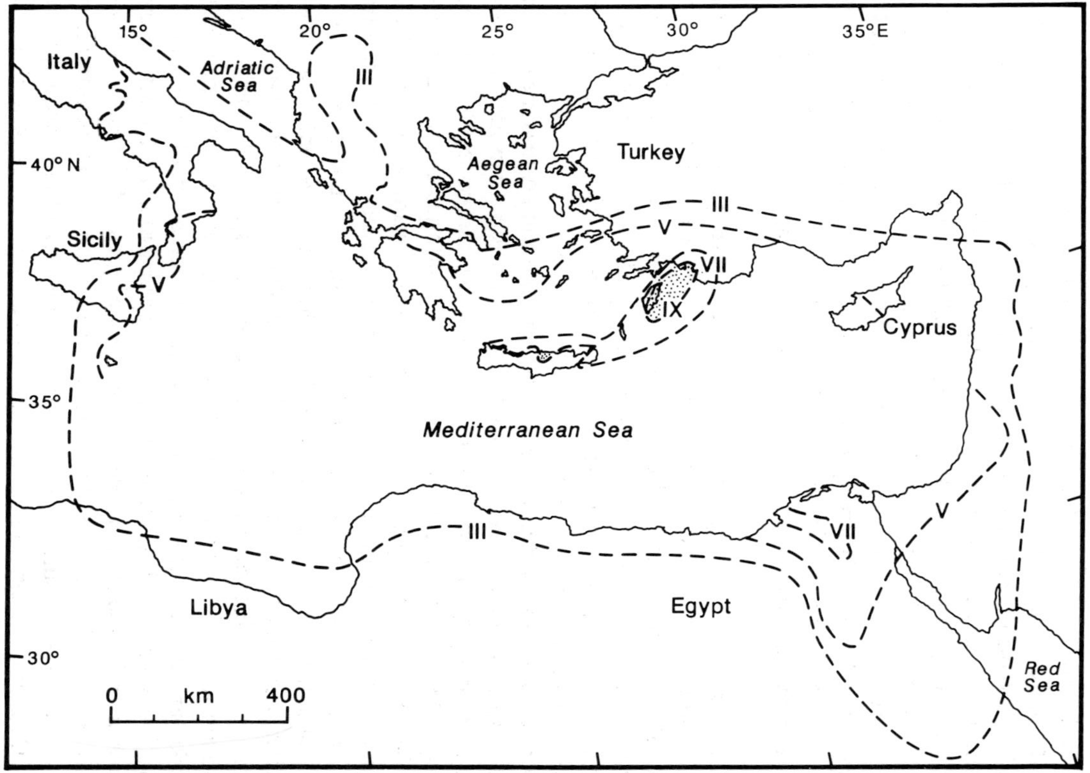

Figure 2.4 The Rhodes earthquake of 26 June 1926: approximate isoseismal map (isoseismal lines encircle areas that experienced equal intensity of ground shaking during an earthquake) (Source: Sieberg, 1932)

In view of the comments made above concerning the greater vulnerability of high-rise structures (especially those on thick sequences of unconsolidated sediment) to low-frequency seismic motions emanating from high-magnitude distant earthquakes, and of the recent earthquake loss experience in Cairo, it is worth contemplating the possible impact of an earthquake of this type upon the developing cities of the eastern Mediterranean today. Since the 1926 earthquake high-rise construction has become prevalent not only in the cities of northern Egypt but also in the coastal cities of Turkey, Syria, Lebanon and Libya. If the failure of the fourteen-storey building in Cairo is taken to be symptomatic of the standards of some high-rise construction in the more congested and inadequately resourced cities of the eastern Mediterranean, then worse may yet be to come as far as earthquake losses in the Nile delta and its wider region are concerned. It is worrying to note that because of the relative lack of high-magnitude intermediate-depth earthquake activity along the Hellenic arc of the south Aegean during the last fifty to sixty years, the probability of at least one such occurrence in the next ten to twenty years is high (Papadopoulos, 1986; Papazachos, 1988).

The importance of public buildings and emergency services

In the above discussion reference was made to alarming failures of public buildings during earthquakes in Third World cities. Buildings that are particularly significant in this context are those of the emergency services (for example hospitals, firestations, policestations), essential life-services (for example water, gas, electricity and telecommunications) and public workplaces and schools. The first two are important because of the need for services of this type to continue to perform beyond an earthquake impact in order to alleviate the suffering of those affected. The latter are significant because of the large amount of human life that can be exposed in buildings of this type on a day-to-day basis.

Considerable public disquiet was voiced in Mexico City after the 1985 earthquake about the large number of public buildings that collapsed. In particular, modern school buildings were among those structures worst affected; the disproportionate amount of damage experienced by school buildings can be judged from the fact that in the immediate aftermath of the earthquake it was estimated that over one million schoolchildren were without a school to attend (Cole, 1985). Similarly, a disproportionate number of other public buildings collapsed or were severely damaged, most notably the fourteen- and twenty-one-storey Law Court buildings at Pino Suarez, south of the city centre. It is perhaps fortunate that the earthquake occurred at 7.17 a.m. (local time), when most people were journeying to work in the city centre and children were on their way to school.

Had the earthquake occurred one or two hours later in the morning, many more of the public buildings that collapsed would have been fully occupied, and the death toll associated with the earthquake would have been very much higher. One can only speculate at the possible effects that such additional human losses would have had upon the social and political stability of the country.

Of equal concern was the failure of a number of major hospitals in the central part of Mexico City during the earthquake. At the National Medical Centre, south of the city centre, eight out of the nine buildings in the complex were damaged. Adjacent to this, the General Hospital was also seriously damaged and one wing collapsed killing most of the staff and patients. The greatest single tragedy, however, was probably the collapse of the twelve-storey Juarez Hospital (again, just south of the centre) which was reduced to the height of a three-storey building, killing as many as one thousand people. It seems inexcusable that in the central part of the city vital emergency services such as these should have been rendered inoperative by the earthquake. Instead of providing an essential resource to help people and society cope with the disaster, many of the buildings in fact proved responsible for some of the heaviest and most sickening losses. Similarly, in the 1992 Erzincan earthquake in Turkey the collapse of essential public buildings of this type served to increase the vulnerability of urban inhabitants of *all* social classes to the earthquake and compounded the severity of earthquake impact upon them. This issue clearly ties in closely with the earlier discussion about the need to focus on the vulnerability of entire Third World cities to natural hazards. The effects of a collapsed hospital or telecommunication station clearly extend to all members of an urban community, from those who require the services to the highly trained personnel who provide them.

Response

The response to the issues identified above clearly has to be twofold. To begin with, more detailed and precise information concerning exposure to natural hazards, such as earthquakes, in Third World urban areas is required. Despite the large amounts of hazard research that have taken place in recent decades, knowledge of the hazard exposure of many Third World cities is still relatively scant. Equally important, much of the knowledge that already exists is difficult to access and seldom presented in formats that enable it to be understood and used by those who, *potentially*, are in a position to enforce and/or promote hazard-conscious development (for example politicians and policy-makers, planners, engineers, insurers, environment and development agencies and educationalists). To some extent, the United Nations' International Decade for Natural Disaster

Reduction (IDNDR) recognises this problem and attempts to address it in its declaration (Lechat, 1990). A number of current research initiatives are also tackling aspects of the problem, including the *Earthquake Hazard Atlas* project (Degg, 1992b; Degg and Doornkamp, 1990, 1992). This is a continuing project, funded by international reinsurers, which seeks to foster a greater awareness of exposure to the earthquake hazard, particularly on the part of certain professional bodies working in the Third World (for example insurers, engineers and planners). The project aims to present essential hazard-exposure data in accessible and usable formats, thereby promoting a more informed management of earthquake hazard and risk by such professions on a day-to-day basis.

Secondly, equally comprehensive measures must be taken to assess and tackle the nature and causes of vulnerability to earthquake hazard in Third World urban areas. In order to be effective, these assessments should recognise from the outset that the problem is often not confined to the poorest urban dwellers but lies in the relative vulnerability of entire Third World cities to the hazard. Over-emphasis on the vulnerability (to earthquakes) of particular social groups within such cities can lead to inaccurate/unbalanced perceptions of hazard exposure and fails to highlight the full scale of the problems faced. It may also serve to induce a blasé attitude among those social groups not singled out as 'vulnerable' and lead to increased fragmentation and alienation within cities. Surely it is only by emphasising the vulnerability of the Third World city in its entirety to earthquake hazard that real pressure can be brought to bear on politicians and governments to tackle the failings of the global economic and political systems responsible. To date, these issues have not been addressed directly by the United Nations' IDNDR.

Conclusions

This chapter has examined a number of factors that influence exposure and vulnerability to earthquake hazard in Third World urban areas. The main conclusion regarding hazard exposure is that the largest Third World cities are, on average, situated in more hazardous parts of the world than those in Western countries. This is the natural environmental context against which studies of human vulnerability ought to be set. It is a consequence not of social or political factors but of the distribution of global tectonic processes.

Set against this, in recent decades the adverse socio-economic and political conditions existing in many Third World regions have often served to place more and more people in urban areas at greater risk of loss from extremes in the natural environment. Owing to inadequacies in the construction of buildings of a wide range of types and uses, and in the provision of essential and secure emergency and support services, this vulnerability often extends across the broad social spectrum of Third World urban dwellers. To single out any one social group as being more vulnerable than another is sometimes misleading and draws attention away from the central issue: that is, the dramatic and unacceptable differences that exist between the vulnerability of Third World cities *en masse* to human losses from earthquakes as compared to their counterparts in Western societies.

Acknowledgements

The author is indebted to The Royal Society of London, Chester College of Higher Education, the University of Liverpool (Board of College Studies Research Liaison Committee) and the London Insurance and Reinsurance Market Association for funding aspects of the research on which this chapter is based; also to Dr Jane Hobson for helpful comments regarding the text at various stages of preparation.

References

Alexander, D., 1991, 'Natural disasters: a framework for research and teaching', *Disasters*, **15** (3), pp. 209–26.

Ambraseys, N.N., 1991, 'Seismicity of Egypt: the earthquakes of 1811, 1847 and 1899' in J. Cosgrove and M. Jones (eds), *Neotectonics and resources*, London, Belhaven Press, pp. 148–57.

Ambraseys, N.N. and Melville, C.P., 1988, 'An analysis of the eastern Mediterranean earthquake of 20 May, 1202' in W.H.K. Lee, H. Meyers and K. Shimazaki (eds), *Historical seismograms and earthquakes of the world*, London, Academic Press Inc, pp. 181–200.

Burton, I., Kates, R.W. and White, G.F., 1978, *The environment as hazard*, New York, Oxford University Press.

Cole, J.P., 1985, *Earthquake in Mexico City: one man's impressions*, Geography Department publication, University of Nottingham.

Degg, M.R., 1986, *The 1985 Mexican earthquake*, London, Reinsurance Offices Association.

—, 1989, 'Earthquake hazard assessment after Mexico (1985)', *Disasters*, **13** (3), pp. 237–46.

—, 1990, 'A database of historical earthquake activity in the Middle East', *Transactions of the Institute of British Geographers, New Series*, **15** (3), pp. 294–307.

—, 1992a, 'Natural disasters: recent trends and future prospects', *Geography*, **77** (3), pp. 198–209.

—, 1992b, 'The R.O.A. Earthquake Hazard Atlas project: recent work from the Middle East' in G. McCall, D. Laming and S. Scott (eds), *Geohazards*, London, Chapman and Hall, pp. 93–104.

—, 1993, 'The 1992 Cairo earthquake: cause, effect and response', *Disasters*, **17** (3), pp. 226–38.

Degg, M.R. and Doornkamp, J.C., 1990, *Earthquake hazard atlas: 2. Egypt*, London, Reinsurance Offices Association.

—, 1992, *Earthquake hazard atlas: 4. Turkey*, London, London Insurance and Reinsurance Market Association.

—, forthcoming, 'Earthquake hazard and urban development in Egypt', *Third World Planning Review.*

Helm, S., 1992, 'Quake puts new cracks in lives of Cairo's poor', *The Independent*, No. 1871, 15 October, p. 14.

Hewitt, K., 1983, 'The idea of calamity in a technocratic age' in K. Hewitt (ed.), *Interpretations of calamity from the viewpoint of human ecology*, The risks and hazards series: 1, London, Allen and Unwin, pp. 3–32.

Kebeasy, R.M., 1990, 'Seismicity' in M. Said (ed.), *The geology of Egypt*, Rotterdam, Balkema, pp. 51–9.

Kolars, J., 1982, 'Earthquake-vulnerable populations in modern Turkey', *The Geographical Review*, **72** (1), pp. 20–35.

Lechat, M.F., 1990, 'The International Decade for Natural Disaster Reduction: background and objectives', *Disasters*, **14** (1), pp. 1–6.

Mitchell, J.K., 1988, 'Confronting natural disasters: an International Decade for Natural Hazard Reduction', *Environment*, **30** (2), pp. 25–9.

Munich Reinsurance Company, 1988, *World map of natural hazards*, Munich, the Munich Reinsurance Company.

O'Keefe, P., Westgate, K. and Wisner, B., 1976, 'Taking the naturalness out of natural disasters', *Nature*, 260, 15 April, pp. 566–7.

Papadopoulos, G.A., 1986, 'Large intermediate depth shocks and volcanic eruptions in the Hellenic arc during 1800–1985', *Physics of the Earth and Planetary Interiors*, **43**, pp. 47–55.

Papazachos, B.C., 1988, 'Active tectonics in the Aegean and surrounding area' in J. Bonnin, M. Cara, A. Cisternas and R. Frantechi (eds), *Seismic hazard in Mediterranean regions*, Dordrecht, Kluwer Academic Publishers, pp. 301–31.

Papazachos, B.C. and Papazachos, C., 1989, *The earthquakes of Greece*, Thessaloniki, Ziti Publications (in Greek with English summaries).

Pearce, F., 1991, 'Acts of God, or acts of man?', *New Scientist*, 1769, 18 May, pp. 20–1.

Pugh, D., 1992, 'Cairo quake gives Islam fresh chance to steal a march', *The Guardian*, 15 October, p. 11.

Sauter, F. and Shah, H.C., 1978, *Estudio de seguro contra terremoto*, San José, Costa Rica Instituto Nacional de Seguros (in Spanish).

Sieberg, A., 1932, 'Erdbebengeographie' in B. Gutenberg, *Handbuch der geophysik*, **4** (6), Borntrager, Berlin, pp. 527–1006 (in German).

Susman, P., O'Keefe, P. and Wisner, B., 1983, 'Global disasters, a radical interpretation', in K. Hewitt (ed.), *Interpretations of calamity from the viewpoint of human ecology*, The risks and hazards series: 1, London, Allen and Unwin, pp. 263–83.

United Nations, 1989, *Prospects of world urbanization, 1988*, Population Studies No. 112, (ST/ESA/SER.A/112), New York United Nations, Department of International Economic and Social Affairs.

White, G.F., 1974, *Natural hazards: local, national and global*, New York, Oxford University Press.

3 Industrial hazards, vulnerability and planning in Third World cities, with reference to Bhopal and Mexico City

Gordon Walker

In 'natural hazard' research the notion of human vulnerability has been usefully applied to the analysis of disasters occurring with increasing frequency and severity in the Third World (Hewitt, 1983; Clarke, 1989). Focusing on vulnerability emphasises that disasters are only created by the impact of extreme physical events on *people* and that levels of impact are intimately related to socio-economic conditions. As Susman *et al.* (1983) state 'Without people there can be no disaster. And poor people are generally more vulnerable than rich ones.'

Vulnerability has been less explicitly explored in relation to man-made technological risks, which as industrialisation progresses are increasingly added to the range of factors already degrading the quality of life and the environment in Third World cities. As widely documented pollution levels in some cities have reached levels comparable or worse than the more heavily industrialised West, leading to a range of serious health problems (Hardoy and Satterthwaite, 1989). Major industrial disasters have also dramatically appeared in developing countries, with casualty figures at levels not seen for many years in the West (Shrivastava, 1987; Inhaber, 1985).

It is the poor who are invariably more susceptible to the impacts of such environmental problems. Low-income housing has been observed to occupy areas where pollution levels are high, where dangerous wastes have been dumped and where water courses are heavily polluted (Main and Williams, Chapter 9 in this volume). This chapter provides an analysis of this phenomenon and the vulnerability associated with it using the particular example of industrial disasters. A detailed examination of two industrial disasters in Bhopal and Mexico City is undertaken, both of which led to major losses of life, high casualties and long-term suffering and hardship for the population living near to the industrial sites concerned. It is these people, the environment they lived in and their susceptibility to the impact of an accident event which provide the focus of the discussion to follow.

Accordingly, little time will be spent on details of the on-site events and safety failures which in each case led to the accidental release of hazardous gases – in the case of Bhopal these have been exhaustively debated elsewhere (Delhi Science Forum, 1985; Jones, 1988; Bowonder, 1987; Shrivastava, 1987). Instead, the analysis of each accident will aim to assess the extent to which the vulnerability of those affected by the accidents contributed to the scale of damage caused, reveal the various dimensions of this vulnerability and trace the factors contributing to its creation. Following the two case histories an evaluation is made of how the threat of industrial disasters may be responded to, focusing in particular on the role of land-use planning in achieving and maintaining separation between hazardous facilities and surrounding populations.

The Bhopal disaster

The disaster which occurred in Bhopal, on the night of 2 December 1984, stemmed from the release of thirty to thirty-five tons of a highly toxic gas (methyl isocyanate) from a Union Carbide (India) Limited (UCIL) pesticide plant. A series of appalling lapses in safety standards meant that what could have been a fairly minor incident, largely contained by safety systems, became a major toxic release into the atmosphere. The impact of the gas cloud was to be horrific: official figures, often disputed as underestimates, list 3,300 dead, 86,000 permanently injured and 60,000 seriously injured or wholly or partially disabled (Bhopal Action Group, 1987). A much larger number of people have been affected directly or indirectly by the immediate and longer-term impacts of the disaster.

While a substantial amount of gas was released and methyl isocyanate is now known to be considerably more toxic than many other gases, other accidents involving toxic releases have occurred at other locations without the same scale of fatalities. For example, releases of 160 tons of ammonia in Blair, USA in 1970 and 90 tons of chlorine in Baton Rouge, USA both resulted in *no* fatalities or casualties (Vilain, 1989). This immediately suggests that factors *other* than the scale and nature of the toxic release were involved in determining the scale of its consequences. By approaching the accident from the point of view of the vulnerability of the population affected a number of such factors are revealed.

Population characteristics

Surrounding Union Carbide was a series of shanty towns or 'colonies'. In places only a road separated these communities from the boundary of the plant and, typical of low-income housing areas, population densities were high, an estimated 25,000 per square kilometre (Marshall, 1987). It was principally these high-density, low-income areas that were affected by the gas cloud (see Figure 3.1). Jones, quotes a study by the Centre for Social Medicine, New Delhi which recorded that:

> Those who died are the poorest. More than half the affected people belong to an income group – about 150 rupees per month – which cannot afford two full

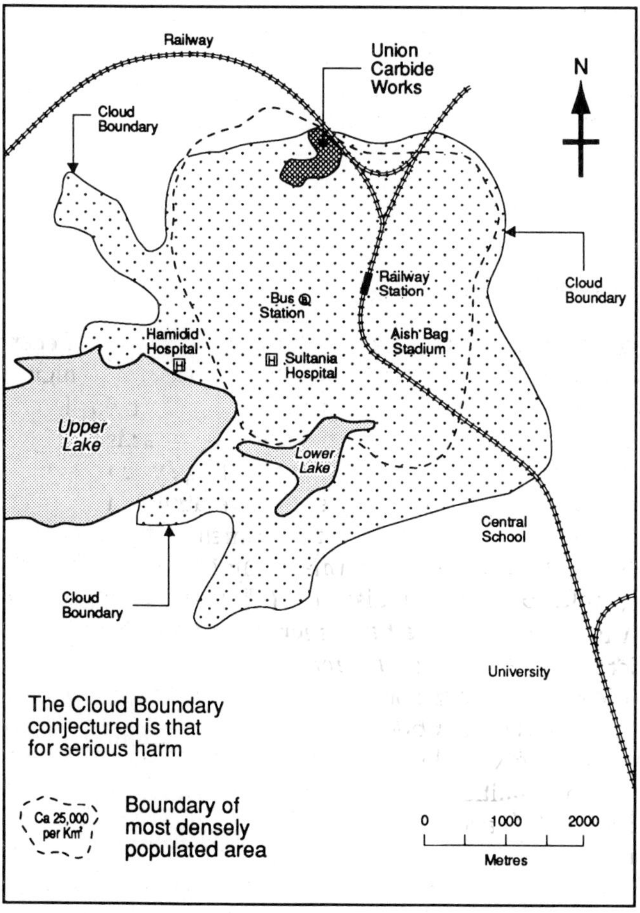

Figure 3.1 Union Carbide and the area affected by the gas cloud in Bhopal (Source: adapted from Marshall, 1987)

meals a day around the year. Those who died were even more disadvantaged than the overall affected population. [Jones, 1988, p. 296]

Lower-density better quality housing was further away from the industrial area and on higher ground better protected against the spread of a gas which is heavier than air (Tachakra, 1989). In crude terms the accident therefore had a high number of people per square kilometre to kill and injure; and those killed and injured were largely the poor.

The creation of this juxtaposition of hazard and high population had taken place over a fifteen-year period, starting with the decision by Union Carbide in 1969 to choose Bhopal as a convenient location for a pesticide-packaging and mixing facility, a relatively simple and safe activity. They were attracted in part by a low-rent and tax agreement for a site on government land to the north of the city close to the railway line; an area which Union Carbide has claimed was relatively empty (Jones, 1988, p. 34; Elkington, 1987, p. 111), but others have argued was already populated when the company arrived (Weir, 1987, p. 36). Rameseshan states that

In 1969 when the licence was first applied for there were at least 50 bungalows around the site of the plant, as well as the Institute of Education and a colony set up by the Bhopal Development Authority. [Rameseshan, 1984b, p. 2109]

From this starting position, changes over the next twenty-five years served to intensify both the potential dangers of the plant and the extent of public exposure to them. At the plant itself expansion from an initial site area of five hectares to thirty-two hectares by 1984 transformed it into a full-scale pesticides production facility. Rather than simply mixing and packaging pesticides, UCIL moved into pesticide production in 1974 using imported raw materials, and then in 1980 began manufacturing the raw materials themselves (including methyl isocyanate) to achieve economies of scale and reduce transport costs. As a processing plant it had become inherently far more hazardous: manufacturing, storing and using a number of toxic chemicals.

The surrounding city also expanded from a population of 102,000 in 1961 to over 671,000 by 1981 – a decennial growth rate of over 70 per cent (Kayastha and Nag, 1989). People moving into the city from rural areas were faced with a familiar combination of poverty, a housing shortage and high land and construction costs. In 1971 a survey found 18,000 squatters living in thirty-three settlements (Garg, 1981). By 1984, 156 slum colonies had appeared in the city (Shrivastava, 1987), including those on marginal land around the Union Carbide plant and other nearby industrial sites, all potential sources of employment and income.

No action was taken by the city authorities to prevent or remove this juxtaposition of hazard and people. Indeed through what seemed to be a progressive policy for dealing with the squatter settlements, a law passed in April 1984 granted ownership rights to existing inhabitants of the shanty towns around the plant (Rameseshan, 1984a). The intention was to consolidate the settlement, legalise occupation and prevent extortion by money lenders and illegal 'landlords'. In the event it also served to consolidate the population's vulnerability to the accident later that year.

Inherent protection

When the accident occurred, the nature of the surrounding settlement also meant that the large number of people living in the area had little inherent protection against the gas cloud. Advice normally given out under European Community requirements to people in Europe living near to plants holding toxic gases is to go indoors, shutting all windows and doors when an accident happens (Gow and Otway, 1990; Lihou, 1985). This gives some protection against the ingress of toxic gas and is generally considered to offer more chance of survival than trying to run away in the open air. Around Union Carbide in poor quality, one-storey and far from airtight squatter homes such protection was unavailable; in the better-off suburbs those living on second or third storeys who shut their doors and windows were apparently little affected (Tachakra, 1989).

Emergency response

Preparation for coping with a major accident affecting the surrounding communities was minimal if non-existent. In the event the response made was chaotic and hopelessly inadequate (Delhi Science Forum, 1985; Rameseshan, 1984a). Information from the plant regarding the release was late, confused and incomplete. Chaos reigned with the police and medical services unaware initially that there had been a release, what the gas was, how dangerous it was and how it should be treated. The local people were equally unaware that such an accident was a possibility and had little idea what they should do when sirens briefly sounded at 1 a.m. when the gas release began, only to be turned off and not sounded again until 4 a.m. In the meantime many ran into rather than away from the gas cloud and did not know that the simple measure of covering mouths with wet cloths could have saved many lives. No organised evacuation was attempted and in any case would have been severely hampered by the density of population, access problems into the squatter settlements, poor communication systems and the low state of preparedness of the emergency services.

Impacts on health and livelihood

The poverty and poor state of health of many of those affected by the gas release meant that they were less able to survive the effects of the gas cloud and more vulnerable to the range of health impacts now documented as resulting from exposure. As the Delhi Science Forum commented in a report on the disaster:

> The vast majority of the most severely affected people are those living in near-starvation conditions. Many of them already suffer from various types of diseases, especially those related to the chest and weakness of body due to overwork and under-nutrition. [Delhi Science Forum, 1985, p. 42]

Already weakened, those inhaling the gas were consequently particularly vulnerable to its impact on the lungs, bronchial tract, upper gastro-intestinal tract and immune system. Disability resulting from the gas cloud also had a greater socio-economic impact on those living on low incomes: they faced 'a shattered economy and shattered health' (Kayastha and Nag, 1989). With work providing the only source of income and survival highly marginal, the loss, or inability to continue work, of the main income earner has had devastating effects on whole family networks (Bhopal Action Group, 1987). Some immediate relief support was provided by the government for the affected communities, but a much vaunted rehabilitation programme, including measures such as landscaping and modernising transport facilities, was strongly criticised as doing little for the gas victims (Das, 1985). Compensation from Union Carbide took four years to agree, was only a seventh of the $3.3 billion claimed by the Indian government and has since been beset by distribution difficulties (Walker, 1990; Hillmore, 1990).

Here it can be argued that much of the delay in receiving compensation, and the inadequacy of the amount agreed can be attributed to the fact that the disaster happened in a developing country and that those affected were poor and politically insignificant. Union Carbide could use the legal problems of agreeing compensation in the Indian courts and the estimated lower cost of an Indian compared to an American life to achieve a 'favourable' settlement (Jones, 1988). The Indian government could afford politically to agree to a low figure (and grant Union Carbide full immunity from prosecution) in order to keep on good terms with multinational investors.

The Mexico City disaster

In November 1984, two weeks before the disaster at Bhopal, a series of fires and explosions at a PEMEX plant in the suburb of San Juan Ixhuatepec, Mexico City, destroyed the plant and led to a major loss of life. Figures are disputed but an immediate death toll of over 500 is generally quoted, with 7,231 officially injured of whom 144 died in hospital (Pietersen, 1985). An estimated 39,000 people were made homeless or evacuated.

Events at the plant are less well documented than for Bhopal, but it has been estimated that a total of 12,000 cubic metres of liquefied petroleum gas (LPG) were involved in the ten-hour conflagration which followed an initial pipeline leak and explosion (Marshall, 1987). While the amount of LPG involved was substantial, again it is the case that similar accidents have happened at other locations with far fewer casualties. An accident at Feyzin in France in 1966, while involving half the amount of LPG, caused only one-thirtieth of the fatalities as nearest housing was more than one kilometre away from the installation. As with Bhopal, in Mexico City the proximity and vulnerability of the communities surrounding the plant were important in determining the scale of the accident consequences.

Population characteristics

Immediately across the railway line which skirted the southern boundary of the site there was an area of very dense low-income housing. Most of the victims lived in this community (see Figure 3.2). PEMEX, a state owned corporation, had set up the plant in 1962 to take LPG from refineries for storage and distribution to wholesalers, LPG being widely used for domestic heating. When first commissioned it was located close to the urban area, near to markets and transport infrastructure, but there was apparently a distance of some 300 metres between the plant and nearest housing (Marshall, 1987).

Over a twenty-five-year period this gap had become much smaller as the city expanded with rapid population growth (Schteingart, 1989), and dense illegal housing spread throughout the suburb whose population swelled to over 40,000 (*Chemical Engineer*, 1985). No planning policies were implemented in order to maintain separation (Chapman, 1984), and it has been claimed that families were even living among the tanks and pipelines within the gas facility itself (*The Guardian*, 1984).

Inherent protection

Housing in the area around the plant was of low quality, and the flimsy construction offered no protection against the explosions which rocked the area flattening all housing within 300 metres of the centre of the site (see Figure 3.2). Fire rapidly spread through the wooden shacks left standing, leading to further devastation. More substantial structures could have offered some protection against blast, fire and scorching radiant heat.

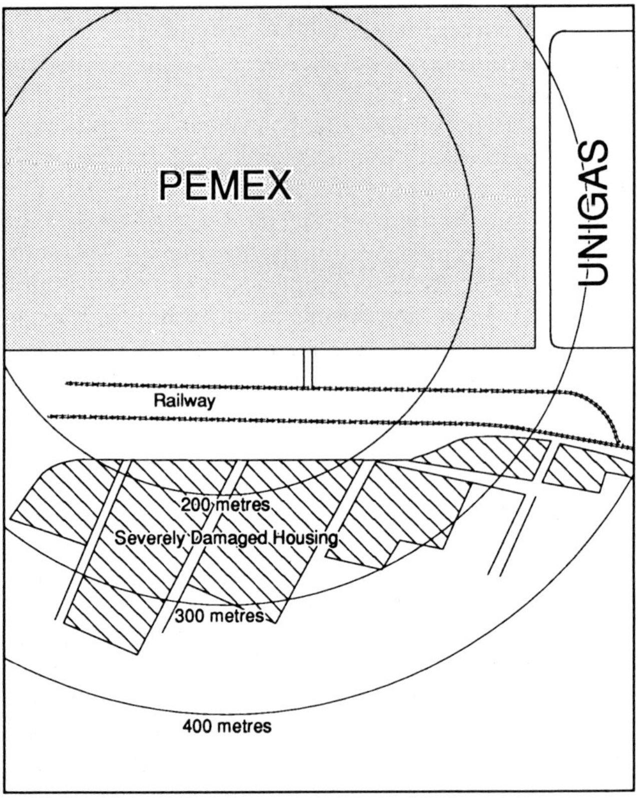

Figure 3.2 The PEMEX site and the affected area (Source: adapted from Marshall, 1987)

Emergency response

Although the accident escalated much more rapidly than at Bhopal, there were in any case no adequate emergency procedures in place to warn or protect the local people. The area was quickly cordoned off, but there was no organised evacuation. Emergency services responded chaotically rather than in accordance with any pre-produced plan (Cullen, 1985). The high population density and poor local communications hindered relief operations, and medical facilities were soon overwhelmed by, and able to do little for, the large number of victims suffering serious burns.

Relief and compensation

While the nature of the accident meant that long-term health problems have been less complex and endemic than in Bhopal, those suffering burns and other injuries were seriously debilitated. The loss or serious injury of wage earners also presented acute problems for families from the area. Compensation was paid out quite rapidly by PEMEX but amounted to only US$ 4 million, and distribution has been a controversial issue. With no reliable records of the population living in the area affected by the accident, the government has been accused of deliberately underplaying the number of fatalities and casualties. It has stuck to an official figure of 499, while locals claim that a further 1,500 went missing (Chapman, 1985). Detailed information apportioning blame for the accident has been hard to come by, with accusations that the government has been keen to cover up a disaster which could not be blamed on foreign multinationals (Chapman, 1984). Claims have also been made that more money has been spent on redeveloping the area than on helping the victims of the disaster.

Dimensions of vulnerability

The two case histories of Bhopal and Mexico City clearly show that human vulnerability played an important part in determining the scale of impact of each accident: in other words that what happened was less important than where it happened and to whom it happened. In both cases the poor predominated in the communities affected. This poverty meant that the areas around both plants were densely populated; that built structures provided little inherent protection against the accident impact; that impoverished people were more susceptible to the impacts of the accidents on their health; that marginal livelihoods were severely disrupted by death and disability; and that relief programmes and compensation agreements could give a low priority to the victims' needs. The immediate emergency response was hampered by the density of settlement and numbers of casualties, and both the companies and public authorities failed to prepare for or inform the public of the possibility of major accident. The vulnerability of the victims thus had a range of dimensions, as summarised in Figure 3.3, each contributing to the level of death, injury and suffering experienced.

Why the poor were predominantly at risk can be attributed to the close and well-documented relationship between industrialisation and urbanisation and the marginalisation of those on low incomes (Susman *et al.*, 1983). Typically, industrial activity locates within or close to urban areas to take advantage of infrastructure, workforce and/or market proximity. Industrial development in turn acts as a magnet for migrants and others seeking employment and income. Land around industrial sites is unattractive to those that can afford to live elsewhere; those who cannot, occupy it (often illegally) because it is empty, available and

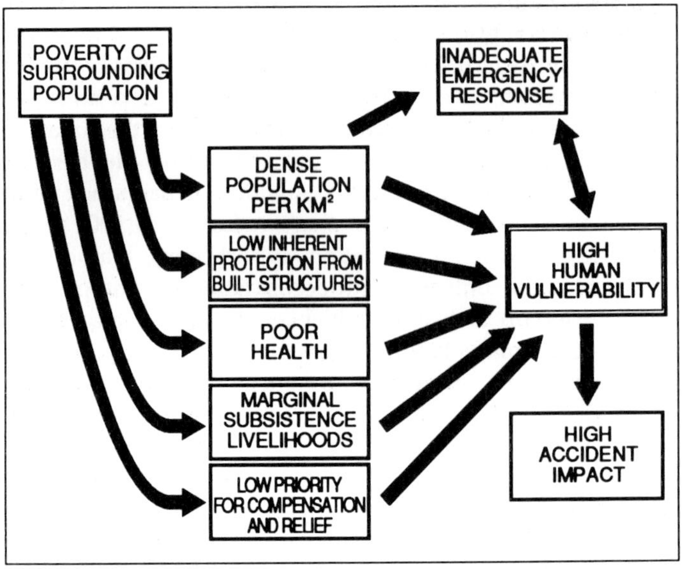

Figure 3.3 Dimensions of vulnerability to industrial disasters

accessible. In Bhopal and Mexico City elements of these processes were evident in the development histories of both plants and the communities around them.

A further factor mentioned in the above accounts was the failure of public authorities to intervene in any way to keep people and hazard apart. After both accidents observers asked why planning powers had not been used to enforce zoning and separation. In Bhopal, it was revealed that the state planning authority had proposed in 1975 that the Union Carbide plant should be moved to a more remote area outside of the city. This policy was opposed by the national government who soon after issued a licence for the plant to expand at its existing location (Bowonder and Linstone, 1987). In Mexico, criticisms were compounded by the identification of a range of other locations where shanty towns surrounded major industrial sites, most notably Pemex's refinery at Azcapotzalco (Chapman, 1984).

Siting and zoning: a realistic solution?

In the face of such criticism Rajiv Gandhi announced two days after the accident at Bhopal that his government was 'planning to stop the construction of factories using lethal gas and toxides near to residential areas' (*The Guardian*, 1984). Subsequently, other observers examining the need for policy responses to industrial hazards in developing countries have identified spatial separation through land use planning as an important element. For example, Kayastha and Nag (1989) state that 'it would be desirable to have green belts around the hazardous factories to keep them away from built up areas'. The Delhi Science Forum (1985) argue that 'all such [hazardous] industries should be removed from urban centres'. The Executive Director of UNEP has called for 'regulations requiring that suitable buffers be placed around factory sites to protect public safety' (Air Pollution Association, 1986).

At face value such statements seem eminently sensible. If people are kept away from hazardous plants and vice versa, vulnerability and hence the impact of accident events will be greatly lessened – particularly where it is likely that nearby development will have a high population density. Address proximity and all the other dimensions of vulnerability to an industrial disaster disappear or are greatly reduced. As is often the case, however, sensible does not equate with straightforward. In practice there are a number of problems and complexities involved in achieving and maintaining the ideal of plant, buffer zone and distant housing.

First of all, it is necessary to distinguish between new and existing plants. Where land uses are well established the inertia inherent in their presence makes it difficult to intervene. Industrial plants represent a substantial physical investment, difficult and costly to move. Even where planning powers are strong and potentially effective, they are in practice rarely used to discontinue a land use as substantial as a chemical plant (Walker, 1984). Where attracting and keeping multinational industry particularly is perceived as a high priority, the simple threat to move investment to other regions or countries may be enough to discourage strong action (Gladwin, 1987). Vaidyanathan identifies a range of interests opposing the removal of existing plants. 'Attempts to move the factories meet strong opposition. Managers complain of the cost, unions of longer journeys to work, and local officials of lost tax revenue' (Vaidyanathan, 1985).

A policy in India to give tax breaks and subsidies to existing firms to encourage them to relocate in designated 'chemical' township zones recognises the costs involved but has been seen as more of a political gesture than an implementable and affordable policy measure.

The alternative of clearing surrounding housing may be equally costly in economic and social terms – and in the experience of attempted squatter clearances ultimately unsuccessful. As Rameseshan (1984a) notes in relation to Bhopal, clearing and resettling the nearby shanty dwellers would have severely disadvantaged those reliant on their location for earning a living in other nearby industries or through a range of self-employed activities. The result, as noted in Weir (1987, p. 104), is that 'Third World leaders have found it unpopular to destroy shanty towns mushrooming around hazard ridden work places. It becomes good politics to let sleeping dogs lie.'

Where new industry is concerned the opportunities for intervention and separation are greater. Areas can be allocated for noxious industry and new plants directed to these (although remoter sites may entail greater infrastructure and transport costs). Once separation is established, it may, however, be easily eroded through urban expansion and population attraction and accumulation – in both Bhopal and Mexico City a gradual but powerfully entrenched process operating over a number of decades. Illegal settlement pays little regard to official zoning in town plans, and leaving an empty 'green belt' may only serve to attract illegal occupation. In response, Batstone (1986) of the World Bank suggests that the dense planting of trees could be used to stop encroachment (in practice chopping these down for fuel would seem a likely response!). Alternatively, actively using land around hazardous plants for activities with low population densities such as warehousing, light industry and transport depots (in combination with effective emergency plans) may provide a more substantial impediment to dense squatter encroachment.

In addition to the more general constraints on planning powers identified above, there are a range of other problems more specific to dealing with industrial hazards which need to be addressed in implementing planning control (Walker, 1991; Petts, 1988). These include the difficulties of identifying sites using hazardous materials; the large range and different characteristics of hazardous materials in common use; the ease with which the use of hazardous materials may be introduced or intensified; and the uncertainties involved in deciding on how much separation is necessary to provide 'adequate' protection.

Faced with the problems outlined above, it would be unrealistic to conclude that using the planning system to separate people and hazards is a universal solution to vulnerability. It can be an incisive and effective means of limiting hazard potential but not when dealing with existing installations, when economic concerns are at a premium, or where land use control is weak and ineffective in the face of illegal development. Political and, more importantly, public pressure can be important in giving safety a priority in land-use decisions, but unfortunately this is more likely to be found after, rather than before, major disasters.

On the other hand such pessimism may partly be tempered by the growing awareness, particularly within multinational industry, that it is in *their* self-interests to achieve and maintain separation. Bhopal was (at least initially) a public-relations disaster for Union Carbide. If separation had significantly reduced casualty levels, the disaster would have made less impact on the world media. Such a perspective may encourage companies themselves to take control of land around hazardous plants and protect against the encroachment of potential victims.

Conclusion

The two case histories discussed in this chapter show that the notion of human vulnerability can be usefully applied to technological hazards. In both Bhopal and Mexico City the poor were predominantly at risk, and this fact provided much of the explanation of why accidental releases of gas became major disasters. While somewhat exceptional in terms of the scale of impact, these disasters stand as realised examples of a widespread potential for disaster at an increasingly large number of plants storing and producing hazardous materials across the developing world (Smith, 1990). The locational conflicts revealed are far from unique, with many communities vulnerable not only to the threat of sudden disasters but also to longer-term pollution problems or 'slow motion Bhopals' (Weir, 1987).

Addressing this vulnerability through separating land uses appears problematic and unreliable. While initiatives, for example, to achieve improved scrutiny and control over siting decisions for new plants are to be welcomed (World Bank, 1985; International Center for Law in Development, 1987; Bowonder and Arvind, 1989), the uncomfortable reality is that people are unlikely to be universally removed or excluded from areas at risk (a conclusion far from unique to developing countries).

One response to this observation would be to argue that more attention should therefore be given to risk reduction at source (by, for example, storing smaller quantities of hazardous materials) and the tighter regulation and inspection of on-site safety standards. Such measures are in any case essential but cannot *guarantee* that disasters will not happen (Perrow, 1984). More attention could also be given to emergency planning, so that when accidents do happen people are promptly evacuated, casualties minimised and injuries more effectively treated.

Alternatively Fazal (1987), among others, has argued for a more radical approach, going beyond mere 'fire fighting' to 'directly confront and fundamentally rethink existing paradigms of development'. Jones (1988) calls for strong action against 'the mass murder that toxic capital entails'. In this respect technological hazards differ from natural hazards in that there is a choice as to whether or not they exist at all and hence whether or not people's vulnerability to them also exists. However, the vulnerability of the poor and industrial disasters will continue to co-exist as long as the dominant view continues to be that 'chemical development is a necessity for economic growth' (Lepkowski, 1987).

References

Air Pollution Association, 1986, *Avoiding and Managing Environmental Damage from Major Industrial Accidents*, Proceedings of APA International Conference, New York, APA.

Batstone, R., 1986, 'Experience in the Application of Hazard Assessment on World Bank Projects' in Air Pollution Association, *Avoiding and Managing Environmental Damage from Major Industrial Accidents*, Proceedings of APA International Conference, New York, APA, pp. 134–53.

Bhopal Action Group, 1987, *No More Bhopals: The Survivors Speak Out*, London, Bhopal Action Group/TICL.

Bowonder, B., 1987, 'An analysis of the Bhopal accident', *Project Appraisal*, **2** (3), pp. 157–68.

Bowonder, B. and Arvind, S.S., 1989, 'Environmental regulations and litigation in India', *Project Appraisal*, **4** (4), pp. 7–22.

Bowonder, B. and Linstone, H.A., 1987, 'Notes on the Bhopal accident: risk analysis and multiple perspectives', *Technological Forecasting and Social Change*, **32**, pp. 183–202.

Chapman, P., 1984, 'Mexico's catalogue of gas disasters', *New Scientist*, 29 November, pp. 3–4.

—, 1985, 'Explosion victims cynical about aid', *The Guardian*, 18 November, p. 4.

Chemical Engineer, 1985, 'Pemex – the forgotten disaster', October, pp. 16–18.

Clarke, J.I., (ed.), 1989, *Population and Disaster*, London, Basil Blackwell.

Cullen, J., 1985, 'What can we learn from Pemex?', *The Chemical Engineer*, October, pp. 21–2.

Das, S.K., 1985, 'Mockery of relief and rehabilitation', *Economic and Political Weekly*, **20** (35), pp. 1679–80.

Delhi Science Forum, 1985, 'Bhopal gas tragedy', *Social Scientist*, **3**, pp. 32–53.

Elkington, J., 1987, *The Green Capitalists*, London, Victor Gollanz.

Fazal, A., 1987 'Foreword' in D. Weir, *The Bhopal Syndrome*, London, Earthscan.

Garg, S.C., 1981, 'Strategy for housing the poor in Bhopal, India: a study of squatters', *Ekistics*, **286**, pp. 79–88.

Gladwin, T.N., 1987, 'Environment, Development and Multinational Enterprise' in C.S. Pearson (ed.), *Multinational Corporations, Environment, and the Third World*, Durham, Duke University Press.

Gow, H. and Otway, H. (eds), 1990, *Communicating with the Public About Major Accident Hazards*, London, Elsevier.

Hardoy, J.E. and Satterthwaite, D., 1989, *Environmental Problems in Third World Cities: a Global Issue Ignored?*, London, International Institute for Environment and Development.

Hewitt, K. (ed.), 1983, *Interpretations of Calamity*, London, Allen and Unwin.

Hillmore, P., 1990, 'Hope falters in King Carbide's city of gasps', *The Observer*, 25 November, p. 14.

Inhaber, H., 1985, 'Risk in developing countries', *Risk Analysis*, **5** (2), p. 87.

International Center for Law in Development, 1987, *Industrial Hazards in a Transnational World*, New York, Council on International and Public Affairs.

Jones, T., 1988, *Corporate Killing: Bhopals will happen*, London, Free Association Books.

Kayastha, S.L. and Nag, P., 1989, 'The Bhopal Disaster' in J.I. Clarke (ed.), *Population and Disaster*, London, Basil Blackwell.

Lepkowski, W., 1987, 'The Disaster at Bhopal – Chemical Safety in the Third World' in C.S. Pearson (ed.), *Multinational Corporations, Environment, and the Third World*, Durham, Duke University Press.

Lihou, D., 1985, 'Why did Bhopal ever happen?', *The Chemical Engineer*, April, pp. 18–19.

Marshall, V.C., 1987, *Major Chemical Hazards*, Chichester, Ellis Horwood.

Perrow, C., 1984, *Normal Accidents: Living with High Risk Technology*, New York, Basic Books.

Petts, J., 1988, 'Planning and hazardous installation control', *Progress in Planning*, **29** (1), pp. 9–75.

Pietersen, C.M., 1985, *Analysis of the LPG incident in San Juan Ixhuatepec, Mexico City, 19th November 1984*, Voorburg.

Rameseshan, R., 1984a, 'Profit against safety', *Economic and Political Weekly*, **19** (51/52), pp. 2147–50.

—, 1984b, 'Government responsibility for Bhopal gas tragedy', *Economic and Political Weekly*, **19** (50), pp. 2109–10.

Schteingart, M., 1989, 'The environmental problems associated with urban development in Mexico City', *Environment and Urbanization*, **1** (1), pp. 40–50.

Shrivastava, P., 1987, *Bhopal: Anatomy of a Crisis*, Cambridge, Massachusetts, Ballinger.

Smith, K.R., 1990, 'The Risk Transition', *International Environmental Affairs*, **2** (3), pp. 227–51.

Susman, P. *et al.*, 1983, 'Global disasters: a radical interpretation' in K. Hewitt (ed.), *Interpretations of Calamity*, London, Allen and Unwin, pp. 263–83.

Tachakra, S.S., 1989, 'The Bhopal disaster: lessons for developing countries', *Environmental Conservation*, **16** (1), pp. 65–6.

The Guardian, 1984, 'The squatters who live at risk', 7 December, p. 12.

Vaidyanathan, A., 1985, 'A sorry technological tale from India', *New Scientist*, 21 February.

Vilain, J., 1989, 'The nature of chemical hazards, their accident potential and consequences' in P. Bourdeau and G. Green (eds), *Methods for Assessing and Reducing Injury from Chemical Accidents*, London, John Wiley, pp. 257–90.

Walker, G.P., 1984, 'Case study: planning applications and major hazards' in J. Petts (ed.), *Major Hazard Installations: Planning Implications and Problems*, Department of Chemical Engineering, Loughborough University of Technology, UK, pp. 66–78.

Walker, G.P., 1990, 'Bhopal: five years on', *Geography*, **75** (2), pp. 158–60.

—, 1991, 'Land use planning and industrial hazards: a role for the European Community', *Land Use Policy*, **8** (3), pp. 227–40.

Weir, D., 1987, *The Bhopal Syndrome*, London, Earthscan.

World Bank, 1985, *Guidelines for Identifying, Analysing and Controlling Major Hazard Installations in Developing Countries*, Washington, DC, World Bank.

4 Continuity through adversity: recovery from the 1960 earthquake in Agadir, Morocco

Hazel Barrett, Howard Fox and Linda Stanier

In February 1960 Agadir, on the coast of southern Morocco, was devastated by an earthquake. Over 12,000 inhabitants lost their lives and large areas of the town were completely destroyed. The recovery of the city as a symbol of the newly emerging state, after political independence in 1956, illustrates well the urban planning response to the earthquake hazard. However, Agadir also presents an example of the continuity of the urban planning process during the twentieth century from early development in the colonial period, through the trauma of the earthquake to the challenges resulting from present-day urban expansion.

The chapter begins by tracing the history and form of the settlement before 1960. The characteristics of the earthquake itself and the resultant damage are analysed. The development of the city since 1960 is described and its re-emergence as a major resort and regional centre discussed.

Urban development of Agadir prior to 1960

The early history of Agadir is not well recorded, but a fishing village named Santa Cruz de Aguer was founded by the Portuguese in 1505 (Péré, 1967). This village was protected by a *kasbah* (fortress) sited on a hill to the north of Oued Tildi, a small stream on the northern side of the Souss Valley. After having been captured by the Saadiens in 1540, the settlement flourished as a port. Its location at the gateway of the very fertile Souss Valley linked it with southern Morocco and Marrakesh and thus connected the port with the trans-Saharan trade routes.

This made the port accessible not only to the rest of Morocco but also eastwards through North Africa and southwards to West Africa. Thus during this period it is not surprising to find the products of a very wide

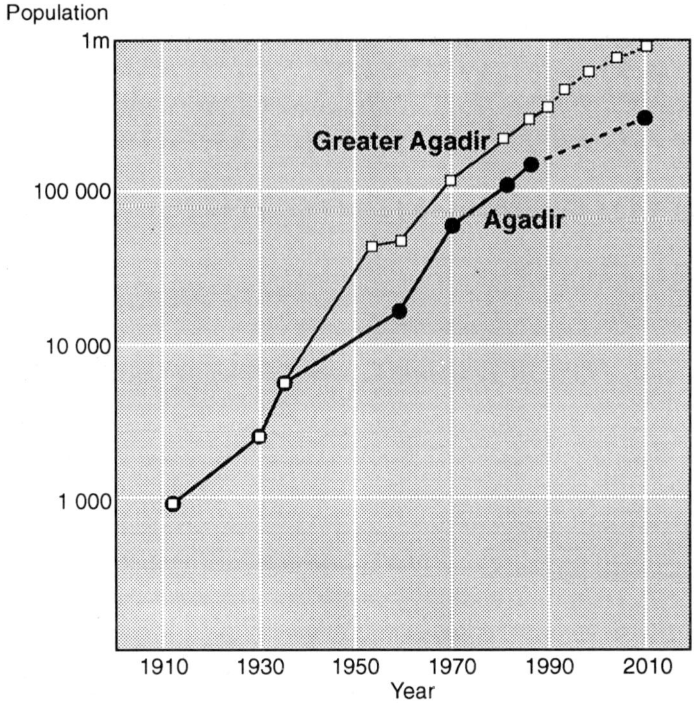

Figure 4.1 Population trends, Agadir, 1910–2010[1] (Sources: Péré, 1967; Direction de la Statistique, 1983)

area exported from the port of Agadir, including dates, almonds, sugar, hides, gum arabic, ostrich feathers and gold. These were principally exported to the European markets of France, Netherlands, Spain and the British Isles. Sugar cane was an important element of this trade but diminished during the seventeenth century with the rise of sugar production in the West Indies.

In 1731 the town was destroyed by an earthquake though this is little documented. Although quickly rebuilt on the same site, the town did not regain its former importance. The general decline in the port's activity was accelerated in 1773 with the development of the port of Essaouira, a little over a hundred kilometres up the coast. This new port successfully captured much of Agadir's trade, and Agadir became known as *'tombe dans l'oubli'* (lost from memory) (Bennani, 1990). Indeed by 1912 the population of the settlement had dwindled to approximately 920 inhabitants (Figure 4.1) and was once again reduced to a small fishing village, with fish being dried, smoked and then sent by caravan to the interior for sale.

The arrival in the town of a French garrison, established in the Founti area beneath the *kasbah*, heralded a revival of the settlement and its port. In 1920 the first French colonists established their wooden buildings in the port area. Speculative development of the Talborjt plateau with more permanent house construction by traders and their employees began in 1925.

However, it was not until 1930 that the official order to open up the Souss valley to development was issued by the French administration, representing a recognition by the colonial authorities of the potential wealth of the Agadir region. The presence of iron, manganese, copper, zinc and lead deposits in the High Atlas Mountains, the agricultural potential of the Souss valley and the rich fishing grounds off the coast all provided the stimulus for the development of Agadir and its port to export these riches. Agadir thus became a municipality in 1930 but lacked the necessary economic and administrative infrastructure. Consequently the first formal development plans for the settlement were produced by the Zeys Commission in 1932 and amended in 1936. These plans identified several new quarters within a greatly enlarged urban area which established the basic structure for the town for the future.

Within these plans existing areas of the town housing the local Moroccan population were to remain. These included the Kasbah itself, Founti and Talborjt. Founti, situated close to the port, was mainly occupied by workers in the fishing and fish-processing industries. Since both Founti and the Talborjt areas were physically restricted and difficult sites to develop, further population expansion consequently required other sites to be opened up. An administrative quarter was developed adjacent to the Talborjt separating the 'native quarters' of the Talborjt from the Ville Nouvelle south of Oued (River) Tildi. This was to house the European population and was developed with villa-style housing at low density. The main industrial quarter had developed at Anza along the coast to the north of the port, but in 1936 a new industrial sector was established to the south-east of the town on flat ground adjacent to Oued Lahouar. Finally, in recognition of the tourist potential of Agadir and this part of Morocco, a tourist sector was planned between the sea and the main coast road. This was to be developed with a reserved beach area and a '*Promenade des Anglais*' as at Nice (Bennani, 1990). Thus in the plan of 1936 the main structural elements of the present urban form of Agadir were established. It would appear that this plan and the early developments of the town took no account of previous earthquake activity, particularly the 1731 earthquake.

With the increasing prosperity of Agadir came rapid growth, with 17,400 people added to the town's population between 1936 and 1952, mainly from migration from the surrounding rural and coastal regions. The inmigration was particularly rapid after 1945, stimulated in part by the employment opportunities offered by the growing fish-processing industry in Agadir and also by the relative decline of oases in such areas as the Draa valley. This dramatic influx of population led to the saturation of areas like Talborjt and Founti, with Talborjt acquiring the characteristics of a traditional *medina* (walled Islamic city). The population densities of Talborjt and Founti were 500 and 426 persons per hectare, respectively, by the early 1940s (Bennani, 1990). In addition, speculative building and *bidonville* (shanty town) developments occurred. Consequently, a renewed planning response was prompted, and in 1946 a plan was produced for the Service de l'Urbanisme by Ecochard. This laid down the principal development projects for the period to 1953 but also made provision for the longer term. A city of 150,000 – 200,000 inhabitants was envisaged by the late 1970s.

The initial developments of this plan included an extension of the Talborjt to the north, the extension of the Ville Nouvelle with the creation of a new sector, 'Extension X', and the development of commercial and administrative buildings and dispersed villa-style housing in a central city area along the main east–west artery. In addition, an attempt was made to address the problem of temporary speculative dwellings. In 1955 the Service de l'Urbanisme estimated that over 11,500 Moroccans lived in a variety of *bidonville* dwellings (Table 4.1), with the village of Yachech in the Tildi Valley housing over half of this population. Three main types of *bidonville* dwelling were described: traditional tents, reed huts and *'trous à rat'* (holes dug in the ground). The Ecochard Plan directed the relocation of these dwellings to two sectors adjacent to the industrial quarters of Anza and south-east Agadir, with twenty-five hectares of the total of thirty-five hectares set aside for this resettlement around the new industrial area of south-east Agadir.

Finally, the plans for a tourist sector put forward in 1932 were reinforced with the allocation of a large area of dunes for tourist developments. During the 1950s Agadir became a thriving and expanding centre

Table 4.1 Number of people resident in *bidonvilles* around Agadir, 1955

Location	Number of inhabitants
Yachech	6,185
Khiam Battoir	2,000
Talborjt	1,000
Anza	900
Carrières Tarpin	632
Carrières Corcos	350
Founti	300
Kasbah	150
Arhesdiz	45
TOTAL	11,562

Source: Bennani, 1990

of industry and tourism. Between 1946 and 1954 the annual number of European visitors to the resort increased from 3,000 to 15,000. Several modern multi-storey hotels were completed during this period to cater for this tourist influx. Activity in the port increased greatly and, by 1960, more than sixty canneries existed for processing seafood, fruit and vegetables. In addition, the regional administration functions envisaged by the plans of the 1930s and 1940s were growing, and by 1960 Agadir had earned itself the title of *'capitale du Sud'* (capital of the south) (Verdugo, 1990). However, on 29 February 1960 an earthquake struck and brought this burgeoning prosperity to an abrupt halt.

Earthquake: 29 February 1960

By 1960 the town of Agadir had extended across a major geological and geomorphological boundary. The Kasbah, Founti and the village of Yachech had all developed in the foothills of the western extremity of the High Atlas Mountains, while the newer parts of the town lay to the southeast in the major synclinal valley occupied by the Souss river. The Tildi valley separated the older and newer parts of the town. This valley follows the line of the Tildi fault in a northeast–southwest direction, separating the Cretaceous rocks of the Atlas zone from the more recent and less well-consolidated alluvial sediments of the Souss valley (Figure 4.2). A second major fault cuts through the Kasbah Hill itself. Hence the geological and geomorphological character of the town's site provides an underlying physical context for the damaging earthquake event of 1960.

Records suggest that slight tremors had been felt in the area during the previous week and a particularly strong shock had occurred earlier in the day, but the main shock itself occurred at 11.41 p.m. on 29 February. The tremor was measured as 5.75 on the Richter scale, a relatively minor earthquake in comparison to other damaging earthquakes. The San Francisco earthquake of 1906 for example was 5,000 times more powerful. Shocks of the magnitude of the Agadir earthquake are expected at some point in the earth's crust on average every few days and usually cause little damage. However, the focus of this earthquake was very shallow and the epicentre was close to the centre of the town (Figure 4.2). Consequently, a considerable portion of the energy released by the tremors was transmitted directly into the buildings of the town. Furthermore, much of the energy was concentrated in a single primary pulse producing greater amplitude of ground motion than would normally have been expected for an earthquake of this size.

These considerations may explain the highly destructive nature of this relatively minor earthquake. However, in addition, the large amount of damage caused is directly attributable to the poor quality and type of building construction commonly found in the town (Table 4.2).

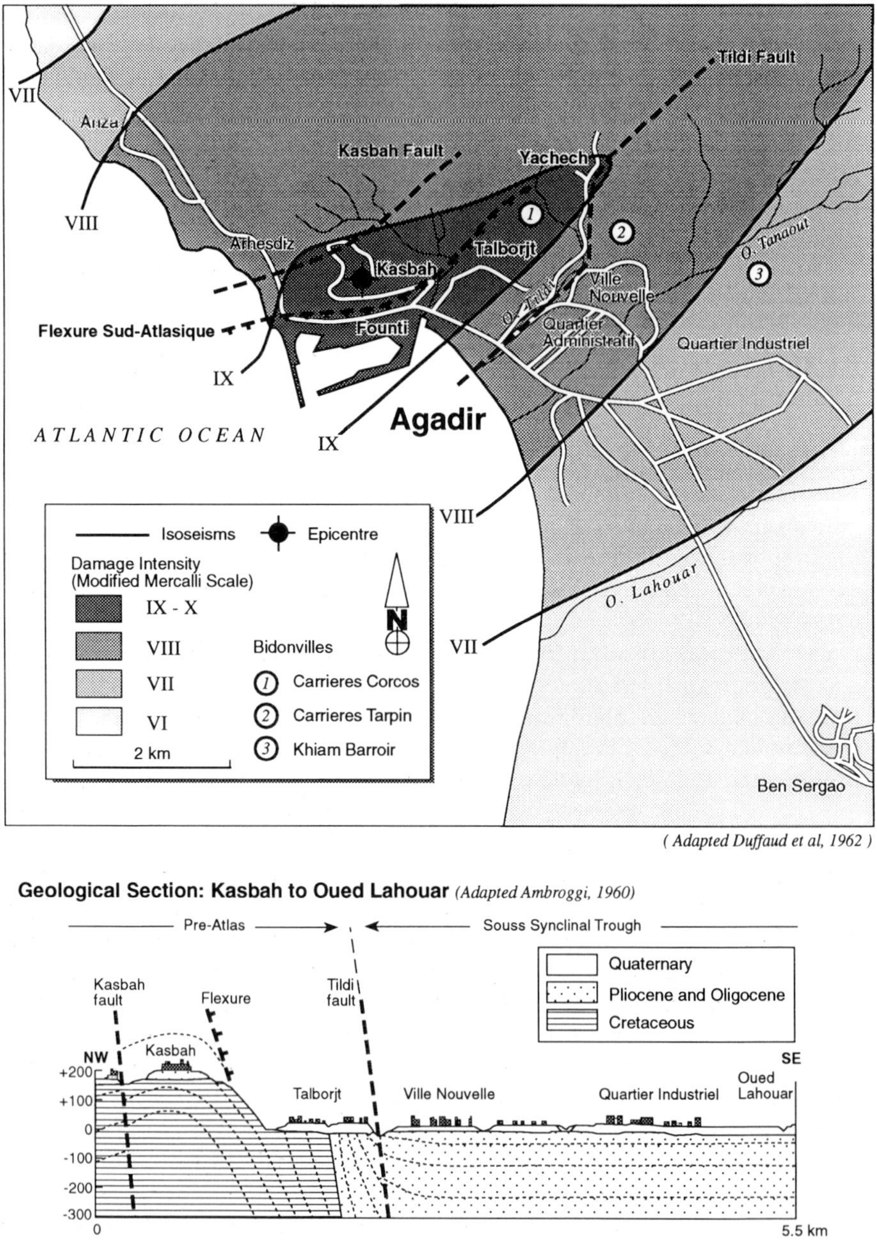

Figure 4.2 Earthquake damage in Agadir and environs, 29 February 1960

Table 4.2 Building damage in districts of Agadir during earthquake, 29 February 1960

District	Percentage of buildings destroyed	Main building type
Anza	0–10	poor quality masonry
Arhesdiz	10–20	" " "
Kasbah	100	" " "
Yachech	100	" " "
Founti	100	" " "
Talborjt	90	two–three-storey masonry
Administrative Zone	70	reinforced concrete & masonry
Ocean Front	60	modern tall offices & hotels
New City	50–60	modern concrete structures
Industrial Zone	20–30	mixed masonry, steel timber & reinforced concrete

Source: after Ambroggi, 1960

Because there had been no *recent* seismic activity in this part of Morocco, no precaution against earthquake damage had been taken in the design and construction of the buildings of the town nor considered in the planning process. Furthermore no serious consideration had been given to the inclusion of earthquake provisions in the building code for any part of Morocco. Hence in the older parts of the settlement poor quality stone masonry structures one to two storeys in height were characteristic and performed badly in the earthquake. They were primarily responsible for a large proportion of the 12,000 fatalities and similar numbers injured which resulted from the earthquake. The districts of Kasbah, Founti and Yachech were close to the epicentre and nearly 100 per cent of their structures were damaged beyond repair (Table 4.2). Most collapsed completely (Figure 4.2). Although newer three- to four-storey masonry structures were to be found in the Talborjt, 90 per cent of buildings were beyond repair. Even the more recently constructed reinforced-concrete buildings in the Ville Nouvelle and the tourist sector failed spectacularly in the earthquake. In several such structures masonry filler walls framed between concrete columns and girders were in fact the principal resistance to lateral forces, and these buildings often collapsed totally.

The spatial distribution of damage is indicated by Figure 4.2 which is based on a modified Mercalli intensity scale.[1] The figure shows the damage assessment for different parts of the town carried out immediately after the earthquake by Ambroggi the Chief Engineering Geologist for the Ministry of Public Works. It can be clearly seen that proximity to the earthquake epicentre was of great significance. Severe damage and loss of life occurred in areas close to the epicentre while districts only two to three kilometres away were little affected. It would appear that the sediments of the Souss valley acted to dampen the shock waves of the earthquake such that only a short distance from the epicentre ground motions were considerably reduced. However, as already noted the inherent earthquake resistance of the buildings in the various districts of the town was also of critical importance in determining the spatial pattern of the damage in the area to the south of the Kasbah fault. Anza and Arhesdiz to the north-west were less severely damaged despite their poor quality building materials owing in part to the geological character of the area.

In his report Ambroggi (1960) warned that the town should not be rebuilt north of the Tildi fault line. He also recommended that south of the river planned reconstruction should pay careful attention to earthquake-proof building techniques. He pointed out that while with greater distance from Oued Tildi such measures were of decreasing importance, if the same epicentre were reactivated with a magnitude of 7 or 8 on the Richter scale, then these areas too might be seriously affected. The significance of this warning was reinforced by slight tremors which occurred in 1989 and by a shock measuring 5.0 on the Richter scale in April 1992.

The townscape suffered no damage as a result of these recent tremors. This confirms the efficacy of the earthquake-proof building techniques recommended by Ambroggi, which were adopted in the reconstruction process, and of the town's spatial planning after 1960.

Reconstruction of Agadir

Planned developments

King Mohammed V visited Agadir on 1 March 1960, the day after the earthquake, in order to survey the magnitude of the disaster. He placed his son, Crown Prince Moulay Hassan, in charge of the rescue operations and signalled that the settlement would be rebuilt as a national priority. The royal patronage is commemorated in a civic sculpture sited in the City Square which quotes the king's solemn pronouncement confirming the intention to reconstruct the settlement: 'The hand of fate may have destroyed the town of Agadir but the people's vision and strength of purpose will ensure her reconstruction.'

A High Commission was set up to oversee and coordinate the reconstruction of the city. Its task initially was to supervise the demolition of irreparable buildings. However, it had broader terms of reference involving powers to annex and purchase land compulsorily, to approve and oversee building authorisations, to ensure the utilisation of earthquake-proof building techniques and to liaise between national and municipal agencies (Lary, 1990). The Commission was empowered to ensure that the reconstruction of the city of Agadir fully utilised the expertise of cooperating national and international architects, developers and engineers. These experts were briefed that reconstruction was to conform to a set of detailed regulations (Verdugo, 1990) affecting the nature of foundations and the form and structure of buildings. The city's plan was to be of low-density layout with land uses strictly zoned and the architectural design was to represent a sympathetic melding of both traditional and modern trends.

The reconstruction programme was funded by a special national 'solidarity' tax, which was levied in July 1960. In total a sum of 350 million dirhams was raised and this was directed to fund the replacement of infrastructure and to repair or reconstruct public buildings and residences. In addition, money was allocated from this fund to provide aid to victims of the earthquake (at the rate of 1,000 dirhams per head of household).

The areas of Founti, Kasbah, Yachech and Talborjt, lying to the north-west of Oued Tildi, had experienced the most severe levels of damage as a result of the earthquake, and the High Commission deemed that rebuilding was not to be allowed on these sites. Those who had occupied these sites at the time of the earthquake were offered a replacement land plot elsewhere in the city on an exchange basis. The site of the new city was designated as a 380-hectare plot to the south-east of Oued Tildi to house a projected population of 50,000. Talborjt was set aside as a memorial garden commemorating those who had lost their lives in the earthquake.

A prioritised rebuilding programme was formulated involving the initial replacement of essential public buildings and public services (hospitals, schools, fire station). Redevelopment of the industrial zone and the tourist sector followed in an attempt to regenerate these aspects of the local economy. Housing and rehousing programmes formed an integral part of the rebuilding process, early developments focusing on Nouveau Talborjt and Les Amicales (Figure 4.3). Royal patronage of the programme and the local municipal authority's determination to rise above the tragedy fuelled the progress of the rebuilding programme. By 1966, only six years after the earthquake, three-quarters of Agadir had been rebuilt and handed over to the municipality (Hicks, 1967).

The new city's urban form encapsulates the harmonious juxtaposition of European-influenced and Islamic indigenous architectural tradition. Wide tree-lined boulevards are fringed by futuristic public buildings, whereas in the residential areas carefully zoned, traditionally styled housing incorporates an inner courtyard forming the heart of the family unit. The new city's functions were to continue to revolve around tourism, port activities, industries (food processing) and administration. Thus the economy of post-earthquake Agadir consolidated those sectors which were successful before 1960.

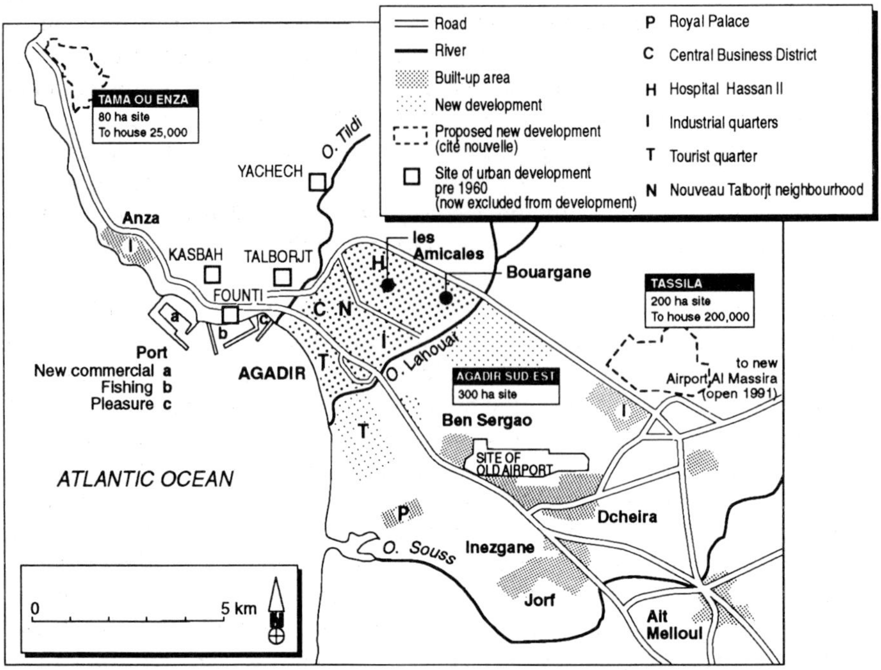

Figure 4.3 The growth of greater Agadir since 1960

Urban development and housing needs

The scale of development in Agadir was rapid and the new city acted as a growth pole in the region, attracting migrants from the Souss valley and beyond lured by the perceived employment opportunities. By 1977, the population of the city had grown to 95,000 (Figure 4.1) and problems reflecting stress on the urban system were apparent. For instance, unemployment levels exceeded national average figures, and shortages of conventional housing had been associated with widespread development of *bidonvilles*. It was estimated in 1977 that 20 per cent of the population of the city lived in shanty town conditions (Barbieri, 1990). In the face of this rapid population increase and its associated *bidonville* developments, a medium- and long-term planning strategy was formulated. The Schéma Directeur de l'Aire Urbaine d'Agadir (SDAUA) was published in 1978, and it included provision for transport, industrial and tourist developments in Agadir. It formulated a planning strategy for the polycentric conurbation of Greater Agadir which incorporated Agadir, Inezgane, Dcheira, Ait Melloul and Tassila (Figure 4.3).

The SDAUA incorporated 'green wedges', zones of forested land, public open space and peri-urban agricultural land uses. The core settlement, Agadir, was viewed as having specialist functions based on fishing, port activities and tourism, whereas Inezgane would constitute the regional administrative and commercial settlement with a renowned regional *souk*/ marketing function. This plan integrated Ait Melloul into the conurbation as the industrial centre. A major feature of the SDAUA was the incorporation of three *cités nouvelles* (new towns) to be constructed as self-help housing areas carefully planned with services, public open space and commercial functions.

The planners in their attempt to create a Moroccan town did not wish to reproduce the urban fabric and associated physical problems found in traditional *medinas* and spontaneous settlements, but they did identify aspects of these settlements that Moroccans valued and then attempted to imitate them in a setting favourable to twentieth-century urban requirements. The plans produced were therefore an attempt to create an urban model adapted to contemporary Moroccan life. Accommodation was to be traditional in style, and interspersed with traditional artisanal, commercial and service use, as found in traditional Moroccan settlements. At the centre of the settlements were to be medical, commercial and spiritual buildings, with schools and sports facilities to be located in peripheral locations. Access, one of the major problems in the traditional *medina*, was to be a top priority and was to be incorporated into the cellular nature of the town. There were to be three grades of road: primary routes (twelve to fourteen metres wide) feeding into secondary roads (eight metres wide), which in turn would feed into a network of tertiary roads (five metres in width).

This system would integrate the settlement, enabling each area to be accessed by vehicular traffic. Each settlement was to contain a social mix, with over 17 per cent of the land area apportioned to affordable low-cost family accommodation and approximately 30 per cent set aside for public gardens (Table 4.3).

In order that the new towns met their target of providing housing for the growing population of Agadir (up to 46 per cent are poor in-migrants), much of the proposed accommodation had to be of the self-help type. To achieve this objective and to maintain the planned character of the towns, the authorities would provide the façade of the buildings but leave the occupant to build their own accommodation to the rear. This made the new housing areas accessible to the vast majority of the population, who had few if any financial resources. It is only by pursuing this strategy that the authorities could hope to clear *bidonville* developments and so keep the urban fabric under control.

Of the three *cités nouvelles* designated (Agadir Sud Est, Tassila and Tama ou Enza), development has taken place thus far only in Agadir Sud Est. To the east of the existing urban area, low-cost family dwelling units, publicly

Table 4.3 Proposed land use in the new towns of Agadir, 1978

Proposed land use	Percentage of urban land area
Low-cost one-family accommodation	10.8
Low-cost multi-family accommodation	6.7
Low-cost multi-family accommodation, artisanal, commercial & service	9.6
Buildings collectively used for artisanal, commercial & service functions	3.3
Villa-style one-family accommodation	13.0
Shops	0.7
Industry	8.7
Sports facilities	13.0
Administration	4.4
Public gardens	29.8
TOTAL	100.0

Source: Bécard *et al.*, 1984, p. 95

subsidised, represent an attempt to allow access to the housing market for poorer families. Planned settlements at Tassila and Tama ou Enza, although first mooted in 1978, have only recently been initiated. Between 1978 and 1985, 16,600 public housing lots were laid out in Agadir, mainly in Agadir Sud Est. Despite this initiative, 18 per cent of Agadir's population (10,500 households) in 1985 still lived in *bidonvilles* (SGDA, 1989). Policies involving rehousing were often frustrated as the resettlement of one section of the household was associated with adult children remaining *in situ* on the *bidonville* site. However, a specific policy of rehousing has been instigated by Agadir's housing department in which named destination areas are allocated to receive *bidonville* residents moving from prescribed originating neighbourhoods (Table 4.4). This rehousing programme has been devised so as to minimise disruption of social networks, to provide basic self-help housing while bearing in mind access to employment. Rehousing is located in areas which suffered least damage in the earthquake of 1960.

Table 4.4 Planned rehousing of *bidonville* residents in Greater Agadir, 1992

Location of existing bidonville	Location of planned recipient neighbourhood
Anza	Hassania or Taddert
Central Agadir	Bouargane and then Tassila
Along Oued Lahouar	Tassila
Inezgane	Forest zone which will be declassified
Dcheira	Jihadla
Ait Melloul	Admine
Ben Sergao	Dcheira

Source: Zaki, 1992

Economic functions

Tourism Prior to the earthquake, Agadir was an important and well-known tourist centre, earning the title of the 'Miami of North Africa' (Clough, 1962). Its top-quality modern hotels attracted many tourists. As early as the Zeys Plans of the 1930s, tourism had been designated one of the four major functions of the city, along with industry and administration. Tourism was to be a major part of Agadir's future development with a tourist sector designated and developed between the sea and the central business district. In this specially planned sector it was envisaged that high-quality hotels would be built in an aesthetically pleasing environment. As a result, Agadir's attraction to visitors from northern Europe increased, particularly after 1945 with the expansion of public air services. Therefore, it is no surprise to find that the re-establishment of the city as a tourist centre of high repute was seen as an important part of the reconstruction of the city following the earthquake.

The provision of tourist accommodation and facilities formed an integral part of the development plans for the Agadir area (Berriane, 1980). A tourist sector was designated fronting the new town in the first redevelopment plan and this has continued to grow south along the bay. The provision of hotel accommodation in this tourist quarter has been spectacular. One of the first hotels to be built after the earthquake was the high-status Club Mediterranée complex built in the mid-1960s. This was followed by a crescendo of hotel construction, encouraged by interest-free government loans and tax exemptions. Between 1978 and 1985 thirty new tourist complexes were opened, increasing the number of available hotel beds from 5,942 to 13,464. The popularity of the resort and confidence in it are demonstrated by continuing investment in the resort. The most recent phase of development includes a new pleasure port which is currently under construction, and work is progressing on the provision of 7,800 hotel and 3,000 apartment places in the area to the south of Oued Lahouar. Strict building controls are enforced to ensure the quality of the environment. The opening of the new airport Al Massira in November 1991 has consolidated this tourist expansion.

This pleasant environment attracted 311,000 visitors in 1982 and 417,000 in 1984. Despite a downturn in the tourist market in 1991 as a result of the Gulf War, figures picked up in 1992 and future plans envisage an increasing role for tourism in the area. A long-term plan for the region proposes the development of three tourist centres, namely Agadir, Tama and Taghazout, followed by developments in Tiznit and Arhoud. Such a programme would provide approximately 50,000 tourist beds in the region and create an additional 50–75,000 jobs. Such plans are dependent on the continued success of the region in attracting tourists. Currently only the plans for Agadir itself are well advanced.

Port The earthquake which destroyed the city of Agadir in 1960 also affected the port. However, the Moroccan Public Works Department estimated that only 5 per cent of the buildings were damaged (Clough, 1962). Most buildings survived and the port remained open throughout the disaster, with the international relief effort using the port to off-load supplies of food and medicines to the stricken town.

Following the earthquake, Agadir's importance as a port, especially for fishing, was reconfirmed. This role is stressed in all the plans put forward for the reconstruction of the town. As a result the port has received a great deal of investment since 1960. Today the port comprises three parts: the traditional fishing port; a new commercial port; and a pleasure port which is currently under construction.

Agadir is reputed to be the largest sardine port in the world. This activity is based principally in the old traditional port which covers an area of seventy-three hectares. Fish other than sardine are also landed, including skate, eels, octopus and tiger shark. There are two fishing fleets using this port, namely the locally owned fleet of traditionally built wooden boats, which fishes daily in inshore waters, and the high-technology deep-water fleet. The deep-water fleet is Moroccan registered but managed by the Japanese. These vessels stay at sea for up to six weeks and process the fish on board, returning to Agadir to off-load the catch and take on new supplies. The existence of such a fleet with foreign interests confirms Agadir's premier role as a fishing port.

In March 1988 a second port, known as the commercial port, was opened in Agadir. Self-contained, it covers an area of 118 hectares and can handle the import and export of over two million tonnes of goods annually. In 1987, 383,075 tonnes of goods were exported from Agadir by sea, but by 1991 this figure had exceeded 600,000 tonnes. In 1991 the principal exports from the port were citrus fruit (61 per cent) followed by minerals (20 per cent) and an increasing proportion of frozen fish. In the same year imports were more diverse, the principal commodities being hydrocarbons, cereals, fertilisers, wood and paper. The opening of the commercial port has confirmed and consolidated Agadir's strategic regional position in the plan for the area and given a stimulus to its industrial sector.

The pleasure port or marina, which is the third component of Agadir port, is currently under construction. The site covers an area of sixteen hectares with 316 planned mooring units. A 7.2-hectare business and tourist zone, complete with hotel facilities, yacht club and sailing school, is also proposed. This development confirms the city's commitment to both its port and tourist sectors and combines both in one project.

In line with stated plans, Agadir port continues to expand. Imports continue to increase and the opening of the commercial port has encouraged a marked growth in exports from the region. Even before the opening of the new port, 17 per cent of Morocco's imports and 18 per cent of its exports came through Agadir (FAO, 1986), and the proportion seems set to increase with the port playing an ever more important role in the development of the region as a whole.

Industry The infrastructure of the industrial zones suffered surprisingly little damage in the earthquake (see Table 4.2). This was perhaps due to the peripheral nature of these sites (Figure 4.2) which meant they were further away from the fault line than residential and other sectors of the town, but also the steel frame structures of many of the buildings performed better under stress than masonry structures.

In the immediate aftermath of the disaster, Agadir's role as a regional industrial centre was reconfirmed and growth continued. However, in the 1978 plan for Greater Agadir, it was decided that Agadir should concentrate on its role as a tourist resort, with industry being re-located. The plan suggested that industry should be moved out of the south-east of Agadir to new sites in either Ait Melloul or Tassila. A 150-hectare site in Ait Melloul was designated an industrial zone, and a new industrial area was identified at Tassila on the road between Agadir and Ait Melloul (Figure 4.3). The industrial role of Anza was confirmed (Abdeslam, 1990). This plan has since been implemented and the older industrial quarter in Agadir is to be evacuated and re-zoned as housing. One of the main reasons for the dispersal of industry from the city of Agadir appears to be fear of beach and sea pollution and its potential effects on tourism. Foreign tour operators have already noted a deterioration in beach quality (De Boysson, 1990, p. 36).

Industry continues to be a major employer in Greater Agadir. The Schéma Directeur du Grand Agadir (1989) projected figures for formal employment by the end of the century of 4 per cent employed in primary industry, principally fishing, 34 per cent in secondary industries, including food-processing, with 62 per cent in the tertiary sector, of which tourist-related employment accounts for 23 per cent.

Conclusion

Agadir today stands as a testament not only to the resilience, determination and ability of its inhabitants to overcome the tragedy of the 1960 earthquake but also to its natural site endowment. The advantages of its location have overridden any perceived or actual earthquake hazard over the centuries. Prior to the tragedy of 1960 no consideration of the potential geological problems of the site had apparently been taken into account in the development of the settlement. The spontaneous pre-colonial settlement as well as the formally planned French town were built without regard to faultlines in the area. The earthquake in February 1960 resulted in much damage and loss of life despite its relatively small magnitude. It appears that proximity to the earthquake epicentre and the quality of building materials largely determined the extent and intensity of damage to the urban fabric.

Following the earthquake, existing town plans were adapted to take account of the earthquake hazard. Areas of 'no build' were designated and more stringent building regulations enforced. As with the pre-1960 town, the reconstructed city was planned with distinct zones including tourist, commercial and industrial sectors. Three new cities were designated and are under construction, two of which are peripheral to the core of the existing settlement and at greater distance from the Kasbah and Tildi fault lines. Within these new cities approximately 30 per cent of land use comprises low-cost family accommodation. Despite this, in-migration and consequent *bidonville* development continues to be an issue for the planning authorities. As with the new cities, *bidonville* residents are being rehoused in peripheral zones within the conurbation.

Agadir has re-emerged since the earthquake as a high-status resort with a broad economic base. Its principal functions and morphology have been confirmed by numerous planning policies up to the present day. As a result, it has become the hub of a thriving polycentric conurbation and acts as the administrative and cultural growth pole of southern Morocco. Thus Agadir provides a good example of the success of continued urban planning in the face of natural adversity.

Notes

1. Population totals are to be regarded as approximate due to the unreliability of census information as well as local boundary changes.

2. For an explanation of the Modified Mercalli Scale of damage intensity, see Degg's chapter above (p. 41).

References

Abdeslam, O., 1990, 'Réflexions critiques sur le schéma directeur de l'aire urbaine d'Agadir 1978–1980', *Rev. Nationale d'Architecture et d'Urbanisme*, special issue to mark the thirtieth anniversary of the earthquake, pp. 16–18.

Ambroggi, R., 1960, *Le séisme d'Agadir 29 février*, Royaume de Maroc, Ministère des Travaux Publics.

Barbieri, N., 1990, 'Trois cités nouvelles pour 100000 inhabitants autour d'Agadir', *Rev. Nationale d'Architecture et d'Urbanisme*, special issue to mark the thirtieth anniversary of the earthquake, pp. 22–6.

Bécard, L. *et al.*, 1984, 'Trois cités nouvelles autour d'Agadir', Paris, *Institut d'arrangement et d'urbanisme de la région d'île de France*, pp. 73,95.

Bennani, C., 1990, 'Agadir: phases historiques du développement d'une capitale régionale', *Rev. Nationale d'Architecture et d'Urbanisme*, special issue to mark the thirtieth anniversary of the earthquake, pp. 39–44.

Berriane, M., 1980, *L'espace touristique Marocain*, Fascicule de Recherche 7, Poitiers, Centre interuniversitaire d'études méditerranéennes.

Clough, R.W., 1962, *The Agadir, Morocco earthquake February 29 1960*, New York, Committee of Structural Steel Producers of American Iron & Steel Institute.

De Boysson, P., 1990, 'Assainissement du grand Agadir', *Rev. Nationale d'Architecture et d'Urbanisme*, special issue to mark the thirtieth anniversary of the earthquake, p. 36.

Direction de la Statistique, 1983, *Population Legate du Maroc*, Casablanca, Ministère du plan et de la formulation des cadres et de la formation profession.

Duffaud, F. *et al.*, 1962, *Le séisme d'Agadir du 29 février 1960*, Notes et mémoires du service géologique 154, Rabat, Edition du Service Géologique.

FAO, 1986, *Annual Report*, New York, United Nations.

Hicks, D.T., 1967, 'Rebuilt Agadir', *Architectural Review*, **142**, pp. 292–300.

Lary, J., 1990, 'La reconstruction d'Agadir', *Rev. Nationale d'Architecture et d'Urbanisme*, special issue to mark the thirtieth anniversary of the earthquake, pp. 7–9.

Péré, M., 1967, 'Agadir; ville nouvelle', *Revue de Géographie du Maroc*, **12**, pp. 43–90.

SDAUA, 1978, *Schéma Directeur de l'Aire Urbaine d'Agadir*, Rabat, Ministère de l'Intérieur Province d'Agadir Cabinet Pinseau.

SDGA, 1989, *Schéma Directeur du Grand Agadir (1989)*, Rabat, Ministère de l'Intér-Province d'Agadir Cabinet Pinseau, 2nd Edition.

Verdugo, C., 1990, 'La séismologie et les villes: Agadir', *Rev. Nationale d'Architecture et d'Urbanisme*, special issue to mark the thirtieth anniversary of the earthquake, pp. 10–15.

Zaki, M., 1992, (Deputy Urban Planner), personal communication, January 1992.

Section 2

Hazardous environments and low-income neighbourhoods

5 Global system, local problems: environmental impacts of transnational corporations along Mexico's northern border

Leslie Sklair

The purpose of this chapter is to show how environmental impacts of transnational corporations along Mexico's northern border engaged in the *maquila* (export-orientated in-bond) industry, illustrate the problematic and dynamic relationship between global economic change and local environmental hazard. The framework of a 'sociology of the global system' (Sklair, 1991) is outlined to conceptualise the ways in which global capitalism creates the conditions for specific types of local and global environmental change. The fact that most Third World and Eastern Europe countries are striving to clamber aboard the 'export-orientated industrialisation' bandwagon suggests that this case has more than local significance.

Sociology of the global system

The 'natural' approach to the global system is state-centrist in the dual sense that it emphasises the role of the state and cognitively privileges the system of nation-states. While not ignoring the nation-state, the approach proposed here offers in addition a conception of the global system based on *transnational practices*. Transnational practices are practices that cross state boundaries but do not originate with states. They are analytically distinguished in three spheres: economic, political and cultural-ideological. In a world largely structured by global capitalism in its various forms, each of these spheres is typically, but not exclusively, characterised by a major institution. The transnational corporation (TNC) is the major locus of transnational economic practices; what I term the transnational capitalist class is the major locus of transnational political practices; and the major locus of transnational cultural-ideological practices is to be found in the culture-ideology of consumerism.

In the case of the maquila industry in Mexico, the dominant economic forces are TNCs, the groups that own, control and service them make up the transnational capitalist class and the culture-ideology that drives the system is thc promise of an 'American-style consumerism' for all those who commit themselves to the global capitalist system.[1]

Mexico: an overview

Mexico has around eighty-five million people, the majority under eighteen years of age. The country occupies about two million square kilometres and its border with the United States stretches two thousand miles from Tijuana on the Pacific to Matamoros on the Gulf of Mexico. The 1990 census shows that the population is highly urbanised (more than two-thirds live in urban areas, rising to about 80 per cent in the northern border states) and that there are several cities of more than one million inhabitants.[2] By 1980, almost two in every five Mexicans lived in one of the twenty-five largest cities (Gilbert and Varley in Gilbert, 1989, p. 17). Mexico City is said to be the largest city in the world, with a population of twenty million and atmospheric conditions hazardous to health eight out of every ten days. Mexico's environment is said to be 'at the breaking point'.[3]

Usually regarded as one of the Latin American NICs (a forgivable misnomer, but it has had a substantial industrial base for many decades), Mexico's 1989 per capita income of US$ 2,010 means that in the World Bank's ranking of 124 countries there were seventy-four poorer countries (World Bank, 1991). In terms of GDP, it is the world's fifteenth largest economy. Predictably, it is a country of stark contrasts between small but growing economic elites and large multi-faceted groups of very poor and struggling urban and rural dwellers.

The housing situation is Mexico is generally acknowledged to be far from satisfactory. In the 1960s and 1970s (a period often labelled the 'economic miracle') the proportions of owner-occupiers in most cities increased substantially (for example, in Mexico City from 22 per cent in 1960 to 53 per cent in 1980; 28 per cent to 52 per cent in Guadalajara; 31 per cent to 68 per cent in Monterrey; 32 per cent to 58 per cent in Ciudad Juarez; 40 per cent to 52 per cent in Tijuana). This was due to many factors, notably land invasions and self-help housing (see Gilbert, 1989). Housing conditions also improved significantly between 1960 and 1980. The proportion of households in the whole country with piped water increased from 32 per cent to 71 per cent and with drainage from 29 per cent to 51 per cent. The comparative figures for the northern border states are 45 per cent to 81 per cent and 35 per cent to 57 per cent (calculated from Lorey, 1990, Table 305).

The decade of the 1980s was disastrous for the Mexican economy. In 1982, the government was forced to default on its huge foreign debt, the peso plummeted from 57 to the US dollar in 1982 to 120 in 1983, 611 in 1986 and over 2,250 in 1988. It is now more stable at around 3,000 (about three New Pesos). In response to these economic traumas Mexican governments, like others, began to put their faith in export-led industrialisation as a development strategy. In the promotion of this strategy, the maquila (or maquiladora) industry (the Mexican version of the export-orientated in-bond industry found now all over the world) fulfilled *symbolic* and *ideological* functions that far exceeded its relatively modest concrete contribution to the Mexican economy. Nevertheless, it is significant that for most of the 1980s the maquila industry was the 'one bright spot' in the Mexican economy in terms of job creation and inward foreign investment (see Sklair, 1993). One important consequence of the growth of the maquila industry is that the *frontera norte* (Northern Border region) of Mexico has been largely incorporated into the global economy, in contrast to the rest of Mexico which has until recently tended to isolate itself from the global economy through its import-substitution policies and its restrictive foreign-investment regime.

However, in the 1980s, a dramatic change of policy came about owing to the concerted efforts of what I have termed the transnational capitalist class, in Mexico and the United States (see Sklair, 1992). The main landmarks along its road to power are Mexico's accession to the GATT in 1986, the liberalisation of inward foreign investment and the negotiations over the Free Trade Agreement with the United States and Canada. The transformation of Mexico from one of the most protectionist and strict foreign investment regimes into what the major institutions of global capitalism disarmingly call an 'open economy' is well under way. As Barkin argues in his challenging critique of this unfolding strategy, 'Mexico's economy has been literally turned inside out' (1990, p. 1).

A North American Free Trade Agreement (NAFTA) was concluded by the representatives of the United States, Mexico and Canada on 12 August 1992 and came into law on 1 January 1994. The NAFTA is likely to introduce at least some features of the maquila industry into the whole of the Mexican economy. Indeed, many commentators argue that the most probable result of the NAFTA will be the 'maquilisation of Mexico'.[4]

The *frontera norte*

'The modern boundary that separates the two nations today is hardly a logical one in terms of ecology, culture, and history' (Herzog, 1990, p. 35). This illogicality encourages us to look at the two-thousand-mile international border vertically as well as horizontally. Horizontally, it divides the richest country in the world from a country that is by no means the poorest but by common consent was in a state of profound economic and deepening sociopolitical crisis for most of the 1980s. However, if we look at the border states of Mexico and the United States vertically – Baja California and California; Sonora and Arizona; Chihuahua and New Mexico/ west Texas; Coahuila/Nuevo Leon and central Texas; Tamaulipas and south Texas – some of the richest communities in Mexico in per capita income terms face some of the poorest communities in the United States. Even the exception of San Diego, a very rich community, is misleading, as the actual border with Mexico is not San Diego proper, which is about fifteen miles from Tijuana, but San Ysidro, a poor mainly Hispanic settlement (see Herzog, 1990, pp. 180ff).

Mexico's *frontera norte* has always been as much connected with the south-west borderlands of the United States as it has with the rest of Mexico, and it has always been considered to be remote from Mexico City in more than a geographical sense. The borderlands are meaningful sociocultural entities in their own right and their isolation from Mexico City and from Washington is sometimes self-inflicted and a true reflection of what the peoples who inhabit them consider to be in their own best interests. This has resulted in a long-standing neglect of these regions by their respective central governments, broken spasmodically by commissions, delegations, reports, flurries of official statements, and then further periods of benign and sometimes not so benign neglect. The history of these events and non-events is, of course, complicated by the successive acts of conquest by which Spain, France and the United States and its various constituent parts acquired large tracts of northern Mexico, successfully pushed the Texas border to the Rio Grande, and split the Californias to the Pacific Ocean.

The United States–Mexican borderlands operate through several politico-administrative systems, each with its own degree of autonomy and institutional structures (see Herzog, 1990 on land use; Waller, 1993 on water). Below the level of the nation-states, the local states and the so-called 'twin cities' that face each other across the border constitute a variety of interlocking, sometimes complementary and sometimes antagonistic, social systems. It is all too easy to imagine that the whole border is characterised by the same set of economic, political, and cultural forces and that any local peculiarities are swamped by these. There is some truth in this, but there are also distinct social formations that give the evolution of the maquila industry rather different characteristics in different areas of settlement along the border (Sklair, 1993, chapters 4–7) particularly in the eight main pairs of border cities in the *frontera norte*: Tijuana/San Diego and Mexicali/ Calexico; Nogales/Nogales; Ciudad Juarez/El Paso; Piedras Negras/Eagle Pass; Nuevo Laredo/Laredo; Reynosa/McAllen and Matamoros/Brownsville (see Figure 5.1).

Before the advent of the maquilas, few of the cities on either side of the

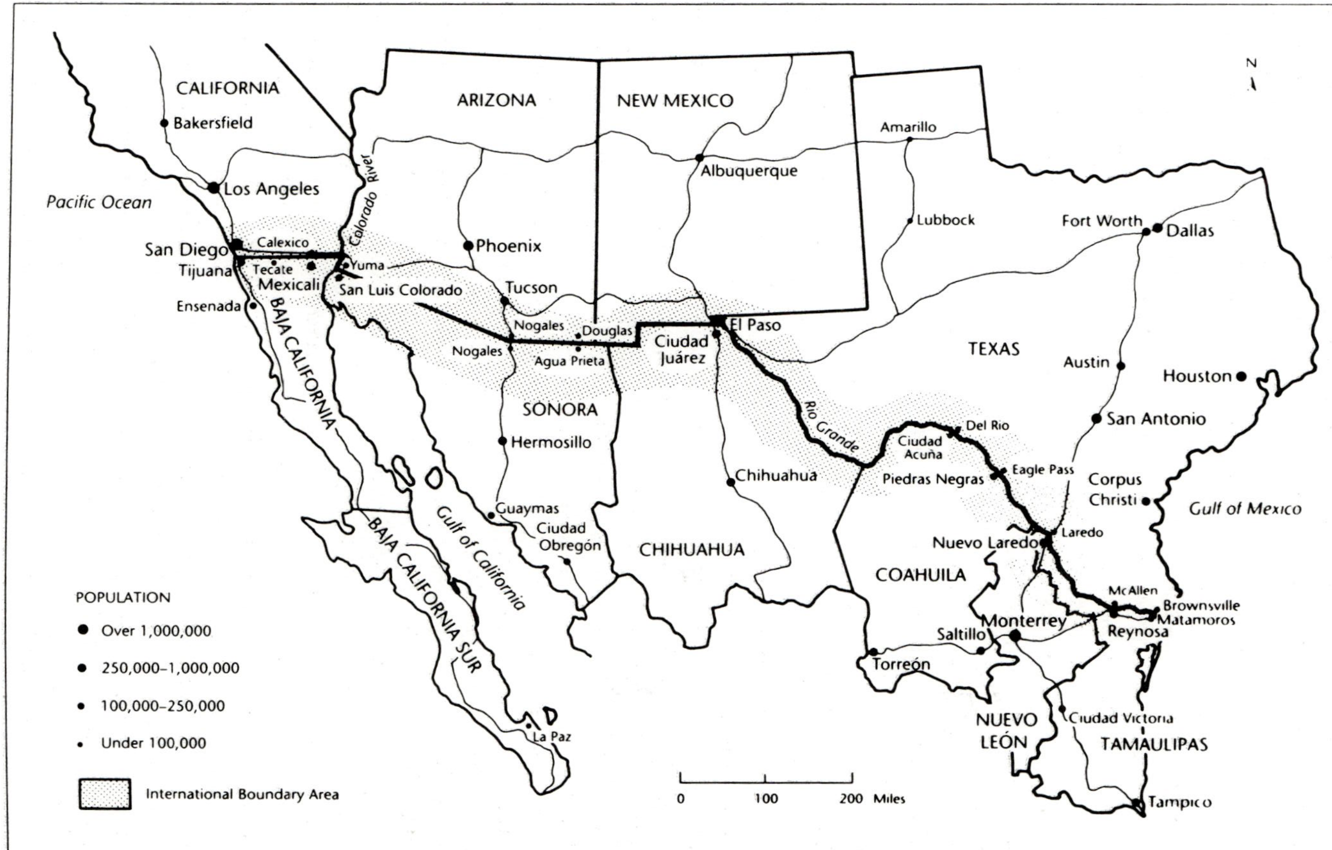

Figure 5.1 Location of the *frontera norte*

border could be characterised as industrial, although the states in which they were located did have some industry. The communities north of the border were for the most part traditionally poor and backward. With the exception of San Diego, some of the least industrialised areas in the United States are to be found along the border (Hansen, 1981). The backwardness of most of the Mexican borderlands had long been recognised. Despite the obvious benefits from the export trade to the United States, there was a good deal of consensus in Mexico that it was the dependence of the *frontera norte* on the rich and powerful neighbour to the north that was at the root of the problem. This was a theme that was to exercise the Mexican government with particular force from the 1960s onwards and was to be instrumental in the creation of a series of administrative vehicles to promote border development (Sklair, 1993, chapter 2).

The maquila industry and border infrastructure

The maquila industry originated in the mid-1960s, when the Mexican government introduced a Border Industrialisation Programme which permitted Mexican and foreign-owned factories to operate along the border duty free on condition that they exported all their products.[5] Incoming machinery, materials and components were held 'in-bond' pending the eventual re-exportation of assembled goods out of Mexico. This is the reason why the maquilas are often termed 'in-bond' industries. The maquila plants, mainly US-owned, were also able to take advantage of US tariff regulations covering the re-import of assembled unfinished goods using US-manufactured components.

From very small beginnings, fewer than one hundred plants and around 4,000 workers in 1967, Mexico now has about 2,000 maquilas, employing half a million workers. The industry employs mainly female operatives, though the proportion of males has been rising in recent years. Three phases of maquila expansion can be identified: gradual expansion up until the United States (and global) recession of 1974–5; rapid expansion from about 1978 until the Mexican collapse of 1982; and mostly rapid expansion into the 1990s, partly owing to under-utilised Mexican plant subcontracting work for the maquila sector. The virtual collapse of Mexican domestic demand during the worst years of the crisis of the 1980s meant that Mexican factory owners were desperate for any type of business and some took on sub-contracts with US corporations under the maquila rules (Sklair, 1993, chapters 3 and 11).

The label 'maquila industry' is something of a misnomer – there are many 'industries' represented in the maquilas of Mexico. The official maquila data base lists twelve sectors. In recent years two, *electronic components* and *auto parts*, have accounted about equally for about half of the

Table 5.1 Growth of urban populations and maquila industry, 1975 to 1990

	1975			1990		
	Urban population (i)	Maquilas	Employees	Urban population	Maquilas	Employees
Tijuana	369	99	7,844	786	469	57,000
Mexicali	429	67	6,324	637	139	22,000
Cd. Juarez	455	86	19,775	840	275	123,000
Reynosa	162	11	1,255	296	60	18,000
Matamoros	200	40	9,778	319	94	37,000
Border total	2,551	413	61,900	4,234	1,644	398,775
Mexico total	51,809	454	67,200	85,472	1,920	446,258

(i) Urban population is in thousands, rounded, and this column is the 1970 figure

Sources: urban populations from Corona (1991); maquila data from INEGI (rounded)

total value added, under half of the workforce and a smaller proportion of the plants. Of the other ten sectors only two (*other manufactures* and *electrical and electronic equipment*) contribute more than 10 per cent of total value added. Although originally designed for the *frontera norte*, since the early 1970s maquilas have been permitted in the interior of the country. The December 1989 'Decree' on the maquila industry strongly encourages maquilas to sell their products in the Mexican domestic market and not simply engage in export-processing, with the implication that maquilas in the interior would be better placed to do this. On the two main indicators of success – jobs created and contribution to Mexican foreign-currency earnings – in 1992 the maquila industry accounted for around 17 per cent of Mexican manufacturing employment and US$ 3 billion.

The advent of the maquila industry has had an enormous impact on urbanisation in the *frontera norte*. Many migrants from all over Mexico have been attracted to the border cities. Although it is difficult to decide the extent to which this is due to the pull of the maquilas and/or the lure of *el norte* (the United States), there is no doubt that rising maquila employment and related economic activity have been responsible for much of the infrastructural strain that is evident in all border cities (see Table 5.1).

Housing

The influx of migrants from all over Mexico to find work in the maquilas and/or across the border in the United States has undoubtedly resulted in severe pressure on the housing sector in most of the *frontera norte* cities. According to the National Maquila Council labour force growth requires

26,000 new houses in Mexico, 20,000 of them on the border. 'The reason for immigration of Mexican border workers to the US is not the lack of employment but the lack of housing, water, electricity, and transportation services' ('Lack of infrastructure', *El Dia*, 22 July 1989). All along the border, from Matamoros in the east[6] to Tijuana in the west,[7] the growth of the maquila industry and the shortage of adequate and affordable housing are linked.

Increasingly, the calls for the maquila industry to accept some responsibility for the general housing and infrastructure crises along the border have grown louder. The NAFTA debate has focused these issues to such an extent that the industry can no longer realistically go on as before. The maquilas do, of course, contribute to Mexican revenues through utility and other costs (power, communications, freight charges, rents and various administrative fees), divided in some unknown proportion between local and federal authorities. Mexican taxes on the maquila industry are minimal and the government has put little funding into *frontera norte* cities for the infrastructural support that the maquila industry requires. The housing subsidy that is collected by the government as part of the burdened wage of maquila workers rarely finds its way back to the border cities.

The thesis that when the Mexican state recycles revenues from the *frontera norte* maquila industry it does so largely to promote the interests of border industrialists and property developers is not outrageous. For example, Herzog shows that most government investment in Tijuana between 1978 and 1982 'has not so much served to improve the quality of life in disadvantaged neighborhoods in the city, as it has to expand economic infrastructure in a few strategic locations'. A meagre 6.7 per cent of the budget in this period went to public housing in what was one of the world's fastest growing cities (Herzog, 1990, pp. 218, 221). The same can be said about most cities along the border.

It seems ironic, but on reflection quite logical, that much of the criticism of infrastructural inadequacies comes from the maquila industry itself. It is in the American Chamber of Commerce's *Maquiladora Newsletter* that we read: 'Employees [in Tijuana] often work in post-industrial factories and return home to pre-industrial conditions' and that the city has not paediatric hospital, roads and bridges are inadequate, there are few technical colleges, transportation is expensive and poor (workers can spend 25 per cent of their wages getting to and from work) and that a Republican State Senator in California is looking for *US funds* to build infrastructure for the maquilas! (Farquharson, 1991, p. 1). Sanchez (1990a) in a more systematic study of living conditions of maquila workers in Tijuana and Nogales concludes that the uncontrolled growth of the maquila industry in these cities has had deleterious effects.

The situation is little better in the other main maquila centres along the border. In 1989 the housing shortfall in Cd. Juarez was between 40,000 and 70,000 units and drainage was non-existent in the outskirts of the city ('Juarez infrastructure atrocious', *El Fronterizo*, 7 May 1989). In the latest of a number of similar critiques, Reygadas (1992) argues that urban and social infrastructure in Chihuahua has not increased sufficiently to meet the demands of the maquila industry, causing severe shortages in housing, health services, nurseries and adult education centres, and leading to increased family breakdown, child neglect, drug addiction and delinquency. While 67 per cent of maquila workers in Chihuahua City (which has a small maquila industry) and residents in the State as a whole own their own homes, only 46 per cent of maquila workers in Juarez (the largest maquila centre) do so. The percentages for renting are 14, 21 and 35 respectively (ibid, Table 5.1).

The situation has reached the point where the authorities in Juarez have been forced to try to curtail maquila development. In 1988, the city froze requests for twelve new maquilas to locate outside industrial parks. Such maquilas, often in residential areas, were said to cause traffic and other problems ('"No" to maquilas', *El Universal de Cuadad Juarez*, 25 November 1988). However, in the desperate search for jobs, it is unlikely that these bans on maquila development will be permanent.

The maquila industry has not stood idly by, though the response has been piecemeal rather than systematic. In Juarez, for example, the maquilas agreed to pay the city 22,000 pesos (under US$ 10) per annum per employee for public works. Sports and medical facilities and child-care centres were among the projects to be funded in this way. In 1988 this raised 1.9 million pesos ('Maquilas contribute to community', *Diario*, 12 February 1989).

The Mexican consul in El Paso, Texas went further and, articulating a growing current of opinion, proposed that maquilas should be obliged to provide certain facilities where they locate. Without discouraging further expansion, he argued, they should be forced to put more into the communities that provide them with their profits ('Infrastructure improved', *Diario*, 2 March 1989). In 1990 the Mexican government introduced a package to reduce taxes, renovate transportation and provide housing and public services to *colonias* along the border. There was talk that this could extend to storage facilities and treatment plants for maquila toxic wastes and that money might be available for factory construction and housing for maquila workers.

Environmental issues in Mexico

C. Richard Bath is not alone when he writes of an 'emerging environmental crisis along the US–Mexico border' (Bath, 1992). Indeed, many scholars and journalists would consider that the crisis, far from emerging, is in full flood. A recent research project of the Monterrey Technical Institute estimated that the border zone generates 9,732 tons of garbage a day and on the Mexican side, 65 per cent of the solid waste (3,286 tons) is dumped in open pits (Herrera, 1992). While there is general agreement that the rapid growth of the maquila industry has intensified environmental degradation along the border, it would be absurd to lay the blame for *all* border environmental problems at the doors of the maquilas. The roots of many current problems, particularly water contamination, air pollution and raw sewage, are longstanding and existed decades before the maquilas arrived (Bath, 1992; Mumme and Nalven, 1988). Nevertheless, the fact that the 1980s saw a definite increase in the sense of urgency in Washington and Mexico City in matters concerning the border environment is at least partly due to the bad reputation that some maquilas were earning for their environmental malpractices.

In the 1980s there were several important bi-national border environmental initiatives. In 1983, Presidents Reagan and de la Madrid signed the La Paz Agreement for Environmental Protection and Improvement and some progress resulted in controlling soil contaminants, cross-border movement of hazardous substances and pollution from copper smelting (leading to the closure of the Phelps Dodge Foundry in Arizona and reduction of emissions from the Cananea smelter in Sonora). La Paz also encouraged the International Boundary and Water Commission (established in 1945 from the 1894 Boundary Commission) to spend $187 million in Tijuana and San Diego, $11.7 million in Nogales and around $40 million in Nuevo Laredo on wastewater treatment facilities. Public health spending along the border increased through local projects of the Pan American Health Organization and there was some binational collaboration on wildlife habitats.

The Mexican government also took some important environmental initiatives of its own in the 1980s. In 1982 administrative responsibility for the environment in Mexico was vested in a new ministry, Secretaria de Desarollo Urbano y Ecologia (Secretariat of Urban Development and Ecology, SEDUE), but it tended to prioritise housing and urban development over environment and ecology (see Mumme and Nalven, 1988). The comprehensive environmental law (*Ley General de Equilibrio Ecologico*), introduced in 1988, gave more power to state and local governments and led to the creation of state and municipal 'ecology commissions'. In 1992, SEDUE was abolished and some of its functions were taken by a larger ministry of social development (Secretaria de Desarrollo Social, SEDESOL), which also housed the politically central Solidaridad welfare and development programme (Wilk, 1992). It remains to be seen whether this change will give environmental issues more or less importance in Mexico.

SEDUE did adopt an apparently proactive policy in the 1980s. The temporary closure of more than 1,500 polluting factories and permanent closure of more than one hundred others (including many maquilas), directing a large part of its budget for regulatory activities; investment of around 3.5 billion dollars in a plan to control auto pollution and improve public transportation in Mexico City; and the permanent closure of Pemex's Azcapotzalco refinery (at a cost of $500 million and 5,000 jobs to eliminate 15 per cent of the capital's industrial emissions) all grabbed headlines in Mexico, the United States and elsewhere.[8] Nevertheless, most of the factories re-opened without significant improvements, enforcement of environmental regulations remains feeble, Mexico City's environment continues to deteriorate as traffic increases[9] and the refinery is being quietly reassembled in Salamanca without significant pollution controls.

The key, of course, is resources of money, trained manpower and effective control systems. For example, there are 6,000 water treatment plants in the United States and only 220 in Mexico. The Mexican environment budget certainly increased sixfold from 1989 to 1991, but at only $39.5 million, this represents $.48 per capita compared with $24.40 per capita in the United States! One hundred extra SEDUE factory inspectors (fifty for the border) brought the total to 225 for the whole of Mexico which is about the same number as for the Los Angeles area. In any case, as the *Los Angeles Times* concludes, for industrial polluters, 'the risk of getting caught is lower than the cost of complying' ('Can Mexico Really Clean Up Its Act?', 17 November 1991).

Mexican industry, in particular, has far to go in environmental upgrading. The Monterrey research project supplied the information contained in Table 5.2 on some of the main industries in Mexico. In the words of the 1991 US General Accounting Office (GAO) report on Mexican environmental regulation: 'Mexico currently lacks sufficient resources and infrastructure to support its rapid population growth and industrialization and to implement and enforce its environmental protection program' (NSIAD-91-227, p. 1). It is significant that the GAO highlights the largely maquila-

Table 5.2 Industries in Mexico and environmental protection

glass: 53 plants, only one with dust equipment
cellulose and paper: 48 plants, two with protective equipment
chemical: 23 companies, two with protective equipment
food and beverage: 64 companies, only two fully comply with the Environmental Law
ceramic and other non-metallic: 35 facilities, only one with blowers/scrubbers
textiles, non-ferrous metals, machinery, rubber: not one has anti-contamination equipment
basic and ferrous metals: only three out of twelve factories have protection devices

Source: Herrera, 1992

driven border industrial growth as a significant factor in Mexico's environmental plight.

Environmental impacts of maquilas

There can be no doubt of the truth of the claim that some maquilas pollute the environment and endanger the health of their workers and the communities in which they are located.[10] Even the highly cautious binational *Integrated Environmental Plan* concedes: 'the total amount of hazardous wastes produced by maquiladoras is still not known and is believed to be significantly higher than the recorded values' (United States Environmental Protection Agency (EPA)/Secretaria de Desarrollo (SEDESOL) Urbano y Ecologia, 1992, pp. 111–20). This is clearly a euphemistic way of saying that no one knows how much toxic waste the maquilas dump illegally, but it is certainly a significant amount.

Environmental hazards originate from both maquila and non-maquila sources, but it is the maquila industry along the border that has been targeted for substantial and at times ferocious criticisms. The tremendous growth of electronics maquilas following relocation from California's Silicon Valley and other centres of the electronics industry where problems of chemical poisoning and toxic wastes are very well documented (see Gassert, 1985) suggests that similar problems for maquila workers and the communities in which maquilas are sited are likely. Electronics is not, of course, the only maquila sector which raises health, safety and environmental problems.

The maquila industry has been fully subject to Mexican pollution laws only since the late 1980s and it may be no coincidence that public anxiety about the environmental hazards posed by the industry intensified around this time. Leslie Kochan, of the Oregon Department of Environmental Quality, researched these issues for the AFL-CIO and her 1989 report is a searing indictment of the record of the maquila industry. While the motives of the AFL-CIO in publishing the report have been questioned, the substance of the report and its sources appear soundly based. The issue is not whether there is environmental hazard along the border, there clearly is, but how to quantify it, and the extent to which the maquila industry is directly and/or indirectly responsible for it.[11] Some of the research cited in Kochan's report is, at best, ambiguous on this latter question and it must also be acknowledged that while some, perhaps many, maquilas are serious polluters, many are not.

There are, of course, other sources of environmental hazard along the border. In a rare recognition of this obvious truth, one of the articles in a *Los Angeles Times* series (see below) identifies five causes of atmospheric pollution in the Juarez–El Paso area originating in Mexico. These are: first, clunkers (old cars imported from the United States to Mexico); second, people making fires; third, fires from Juarez dumps (the city has no proper sewerage system); fourth, the brick factories of Juarez; and fifth, the maquilas ('On Texas Border, Outlook for Air Quality Is Murky', *Los Angeles Times*, 20 November, 1991).[12]

Nevertheless, that the maquila industry is responsible for some types of environmental hazard (for example, pollution of drinking water, illegal dumping of toxic wastes and inadequate protection of the workforce) can hardly be denied. Typical cases are of toxic-waste runoffs from maquilas into residential areas (see below) and hazardous airborne effluents, for example in the 'Barreal' area of Cuadad Juarez where a local maquila was polluting the atmosphere with glass-fibre particles and the carcinogenic, tetrachloroethylene. Complaints filed with SEDUE and other agencies were ignored (see *El Fronterizo*, 7 May 1989).

Several institutions, including the premier Mexican border research centre, El Colegio de la Frontera Norte (COLEF) and the United States' EPA have documented the huge discrepancy between the large amounts of toxic materials entering Mexico for use in the maquilas and the small amounts of toxic wastes returning, as required by law, to the USA. Juarez maquilas are said to dump most of their toxic wastes in the Rio Grande. One COLEF survey showed that only twenty out of 772 border maquilas reported returning toxic wastes to the United States, though 87 per cent used toxic materials. A common practice, reportedly, is for maquilas to sell old chemical drums to employees who sometimes use them for water storage. The media are beginning to link birth deformities and high rates of cancer along the border with toxic wastes.[13]

There is little doubt that some US manufacturers have established maquilas in order to escape strict US environmental regulations, including expensive toxic-waste regulations. For example, some furniture manufacturers from Los Angeles have established maquilas in Baja California to escape legislation on solvent emissions (US General Accounting Office, NSIAD-91-227). Since 1988, when the new Mexican Environmental Law came into force, there has been general agreement between supporters and opponents of the maquila industry that there are too few inspectors and basic tools to enforce the law along the border, let alone in the rest of Mexico (Lowery, 1990; Kochan, 1989). Despite recent reinforcement of SEDUE, much of the opposition to NAFTA between Mexico and the United States has focused on the widespread belief that Mexico will not be able to enforce its current environmental legislation properly.[14] It is notable that maquila spokespersons have become increasingly conscious of the damage that a bad environmental image has inflicted on the industry.[15]

As noted above, the La Paz Agreement of 1983 acknowledged that there were some adverse environmental impacts of the maquila industry, but these were not considered at that time to be very serious, nor was the connection between the growth of the maquilas and the rapidly deteriorating infrastructural situation along the border very obvious. In the context of the devastating economic crisis from 1982 until the end of the decade, it is hardly surprising that the promise of maquila jobs and foreign-currency earnings were uppermost in the minds of Mexican policy-makers. By 1992, however, it was no exaggeration to claim that the environmental devastation that the maquila industry has brought to parts of the border is firmly implanted in the consciousness of a large and growing sector of the population in Mexico, the United States and elsewhere. What happened?

The answer is, clearly, the North American Free Trade Agreement (NAFTA) and the success that its opponents had in tainting the whole industry with the appalling environmental record of some maquilas. This is a case of an environmental movement winning the battle but losing the war! (see Sklair, 1992a). In 1991 and 1992, all the major US newspapers and TV channels ran stories (often virtually identical) on the hazards that the maquilas had brought and were continuing to bring to the US–Mexican border. There were two clear implications of these stories. First, with rising consciousness of *global* environmental degradation, reinforced by all the publicity given to the Rio Earth Summit in June 1992, the health and welfare of *Americans north of the Mexican border* were directly threatened by what was going on *south of the border*.[16] Second, the NAFTA would most likely increase the numbers of US factories operating in Mexico and without strong safeguards which on past performance the Mexican authorities could not guarantee, environmental degradation was bound to get worse.

Examples from four of the most influential US newspapers will illustrate these points:

(i) In 1991 the *Los Angeles Times*, probably the most important paper in the United States with a significant border readership, printed a four-part series under the title 'The Free-Trade dilemma: the environmental costs of a U.S.-Mexico pact' (17–21 November). Headlines on the various stories in the series make the point vividly. 'Mexico: plan for Free-Trade pact raises fears over environmental controls' (17 November); 'Foreign-Owned Companies Add to Mexico's Pollution' (18 November);[17] 'Firms find a haven from U.S. environmental rules' (19 November); 'Mexicans Fear Plant Could Cause "Next Bhopal"' (20 November).[18] In each of these articles maquilas were clearly identified as polluters and responsible for toxic hazards.

(ii) *The Washington Post* published many similar stories. Edward Cody, reporting from the border, wrote in February 1992 that the Tomatera well in Nogales symbolizes 'a legacy of filth and pollution . . . in a string of dingy factory towns, industrial parks and slums from Matamoros on the Gulf of Mexico to Tijuana on the Pacific Ocean' – a result of the maquila industry's uncontrolled growth. He identifies specific sites, such as the Nogales (Sonora) wash and the Otay Mesa (Tijuana) runoff where maquila toxic wastes endanger the lives of neighbouring communities. In Matamoros, research on birth defects in Taumalipas and South Texas found xylene levels downstream from maquilas at 52,700 times the permitted US standard for drinking water. He quotes the president of the Environment Committee of the Mexican National Maquila Council as asserting that US firms dump 260,000 tons of toxic wastes per annum illegally in Mexico. An agreement of 1986 obliged maquilas to return their toxic wastes to the United States, but a GAO study estimated that of 1,449 border maquilas, only the 446 registered as toxic-waste producers could be monitored ('Expanding waste line along Mexico's border', *Washington Post*, 17 February 1992, pp. A1, A34).

(iii) In the *Wall Street Journal*, Senator Metzenbaum (D-Ohio), identifies a 'growing environmental nightmare on the border' and charges that in the maquilas 'environmental degradation, the working standards and the living conditions are truly unbelievable' ('Bush's Mexico trade plan faces tough road in Senate', 8 May 1991).

(iv) Robert Suro, writing in the *New York Times* in 1991, points out that maquila environmental issues have become a 'lightning rod' for anti-NAFTA forces. Suro correctly identifies a key issue in the debate when he quotes Roberto Sanchez of COLEF in Tijuana: 'They [US environmentalists] want to save the planet, but it is not the same planet on their side of the border as it is on ours. They can afford to defend the environment for its own sake. Our people need to use the environment to survive' ('In Search of a Trade Pact with the Environment in Mind', 14 April 1991).

Newspapers from border communities also repeatedly draw attention to environmental impacts of maquila production along the border. In San Diego, the local paper highlighted a long list of hazards. From Matamoros came familiar reports of maquila toxic wastes and explosions and babies born with neural-tube defects. Pollution of drinking and household water is directly linked to the growth of the maquilas and the consequent overwhelming of municipal infrastructure in Laredo, where the authorities counted coliform bacteria in the Rio Grande at 200 times the recommended maximum for swimming. Hepatitis in San Elizario (near Juarez–El Paso) is five times the national average. When University of Arizona researchers tested public wells in Nogales, Mexico, they found four industrial solvent chemicals at up to five times acceptable levels. One well was closed down but quietly re-opened. Less than half of Baja California's maquilas are in full compliance with SEDUE, whose lack of openness fuels alarm on both sides of the border. The Quimica Organica chemical plant in Mexicali leaked hydrochloric acid in January 1992, and 10,000 locals were evacuated. Violent demonstrations followed and SEDUE was forced to close the plant, but the community was given no information. The plant is to relocate somewhere else in Mexico. In Mesa de Otay maquila park in Tijuana, high levels of lead, chrome, copper, cadmium and zinc run off into Colonia Chilancingo endangering 25,000 people, who have been waiting since 1983 for SEDUE to act ('A border boom has its ugly side: Toxins menace both U.S., Mexican families', *San Diego Union-Tribune*, 16 February 1992, pp. A-1, A-8).[19]

In El Paso, captions for two newspaper pictures eloquently illustrate the concerns raised by the uncontrolled growth of the maquilas. In the first, 'Two women, one holding her breath, hurried across a shallow, feces-laden canal to catch a bus after a shift at a Juarez maquiladora'; in the second, 'Empty barrels, many of which once contained toxic or hazardous chemicals [from maquilas], are sold at a stand . . . in Juarez. The barrels are often used to store water for drinking and bathing' ('Border environment plan "inadequate". Key health issues ignored, experts charge', *El Paso Times*, 20 February 1992, pp. 1–2A).

In Arizona, the alarming testimony of a business consultant to a government committee in Washington is reported to the effect that in many Mexican industrial plants pollution control systems for hazardous fumes are 'operating at a fraction of their capacity or not being utilized at all' ('Mexican plants often hazardous, U.S. panel told', *Arizona Daily Star*, 1 October 1991).

To date, the response of the maquila industry to this woeful record of environmental irresponsibility has been outstanding on paper but less than impressive in practice. At the Environmental Protection Agency's First Annual Border Environmental Assembly, attended by Colosio and Reilly, the respective heads of SEDESOL and EPA, an EPA spokesman (in the words of the correspondent of the major maquila industry trade magazine) 'boasted a handful of maquiladoras put up $15 million toward funding a sewage plant for Juarez, and 239 maquiladoras pledged to guarantee the bond'. However, the General Electric executive who was organising the scheme quickly made clear that the plan was not yet finalised and that the idea was for each maquila to pay $200 per employee in return for $1,000 per year for ten years off its sewerage bills! (Lowery, 1992, pp. 48–50). While the transnational maquila class has been talking a great deal about improving its environmental record over the last few years, particularly while defending itself from the charges widely broadcast during the ongoing NAFTA debate, there is little evidence that the industry is prepared to face the financial and social consequences of its practices and its evolution. In the last resort, it would be cheaper and more economically 'rational' to move to locations with less troublesome environmental regulations. The maquilas know this, the Mexican government knows this and the communities which depend on the industry for some measure of prosperity (or, at least, employment) also know this.

Conclusion

The tendency to transnationalisation of the global capitalist system, notable since the 1950s, accelerated dramatically in the 1980s and 1990s. One consequence of this has been the consolidation of a multitude of transnational corporations engaged in manufacturing, often of the export-processing variety, in low-wage countries. While it is not certain that these are *typically* polluting and hazardous enterprises (Leonard, 1988), the evidence marshalled in this chapter demonstrates that along the Mexican–US border, there are indeed very many TNCs and associated factories that are responsible for major environmental hazards. This is closely connected with the stress that uncontrolled industrialisation and urbanisation have put on the infrastructure of the cities of the *frontera norte*. This analysis of the maquila industry confirms the accuracy of Barkin's view of the 'belated recognition that many solutions to environmental problems are at variance with the prevailing pattern of concentrated, export-oriented development' (Barkin, 1990, p. 47).

As I argue at length in *Sociology of the Global System* (Sklair, 1991), the key public mechanism that global capitalism uses to achieve its ends in the present era is the 'culture-ideology of consumerism'. The fundamental rationale for the almost universal acceptance of the maquila industry and latterly NAFTA in Mexico, despite the proven record of environmental degradation of the one and the other's prospect of more, is that they promise a 'US standard of living' for every Mexican. This, indeed, is the lure of global capitalism for all the underprivileged in the Third World and Eastern Europe, not forgetting those in the First World. The experience of the *frontera norte* should make even the most committed 'free-trader' think twice about the global and local environmental consequences of bringing the consumerist fruits of capitalism to all those who labour to produce its profits. And we have hardly even begun to discuss the resource implications of the capitalist global project!

Notes

1. This is an extremely abbreviated account of a much more extensive argument, for which see Sklair (1991 and 1992).
2. This is deliberately vague owing to the significant undercounting of urban populations, acknowledged by most specialists. For example, the official (corrected)

population of Tijuana in 1990 was 785,870, but most informed estimates put it at well over one million (see Corona, 1991, Table 3; Herzog, 1990, pp. 106–15).

3. The quote is from Mumme (1992, p. 123). There is a considerable literature on environmental degradation in Mexico. See Barry (1992, pp. 265–77) for a brief and lively introduction and Ortiz Monasterio (1987) for more extended treatment.

4. See, in general, Sklair (1992a and 1993, chapter 11). NAFTA has generated a large number of publications, for example, Rey Romay (1992); Reynolds *et al.* (1991); and a flood of newspaper and periodical comment.

5. Data on the maquila industry are published regularly by Instituto Nacional de Estadistica Geografia e Informatica (INEGI). Sklair (1993), from which this chapter borrows, reproduces and analyses selected maquila data.

6. 'Housing shortage for maquila workers' (*El Fronterizo*, 23 March 1990) reports that the housing crisis in Matamoros is so serious that maquilas are experiencing labour shortages.

7. Farquharson (1991) reports that the housing shortage in Tijuana is estimated at 100,000 units, and Sanchez (1990a, Table 2) that 50 per cent of households lack drains and over 30 per cent piped water.

8. For example, 'Environmental problems: a glimmer of hope', *Financial Times*, 25 October 1990, P. VII (Mexico Supplement). The first 'environmental diplomat' was appointed US environmental attaché in Mexico City ('Pollution fighter on U.S. embassy roster in Mexico', *Washington Times*, 8 January 1991, p. 5A.

9. In an imaginative gesture, each private car is banned from city streets for one day a week. Large numbers of Mexico City dwellers evaded the law by buying extra cars! (see Mumme, 1992).

10. Worker safety in the maquilas has had less attention than other environmental issues. Tragedies such as the death of Elizabeth Macias's son in a preventable accident in a Ford maquila in Juarez only achieved general publicity when his mother testified before a Senate Committee against NAFTA ('Mexican trade pact held inimical to tough safety regulations', Associated Press, 23 April 1991). For two rather different approaches to health and safety in the maquilas see Moure-Eraso *et al.* (1991) and Guendelman and Jasis (forthcoming).

11. For a good discussion of the main issues, see Sanchez (1990), Perry *et al.* (1990), Rich (1992) and the excellent video on Stepan Chemicals in Matamoros (Day, 1992).

12. For a rather more hopeful account of the struggle to improve air quality in Juarez, see Wilk (1992). Maquila supporters become understandably agitated when blamed for all environmental hazards along the border. Nevertheless, many seem unable or unwilling to grasp the connections between uncontrolled maquila growth and hazards due to infrastructural inadequacies, a central argument of this chapter. Susan Lowery, who writes for various maquila publications, is a notable exception.

13. See, for example, 'Disfigured children, distressed parents; suit filed against US firm over alleged effects of pollution in Mexican border region' (*Washington Post*, 15 October 1991) on the 'Mallory children', deformed due to toxics from a now closed maquila and the *Los Angeles Times'* series of November 1991, discussed below. The research of COLEF's Roberto Sanchez has been influential here.

14. Hearings on the SEDUE/EPA Draft Border Environmental Plan held in nine US cities attracted over 150 groups and individuals and a torrent of criticism (Kelly, 1991). For a Mexican view see 'De alta peligrosidad, cerca de 400 industrias de EU en la frontera', (*El Financiero*, 3 October 1991, p. 6A). This led to an improved version, US EPA/SEDUE (1992), which silenced some but by no means all the critics (Rich, 1992). This is now the basis for US–Mexican border environmental planning.

15. Alejandro Bustamante, as President of the National Maquila Council, called on

the government to regulate maquilas 'so that they will no longer cause pollution problems for the surrounding population. There have already been numerous complaints about this' (*UnomasUno*, 11 March 1989). Maquila magazines now have regular environment features and most local maquila associations on both sides of the border have environmental committees.

16. I have attempted to theorise the sociology of global environmental change in Sklair (forthcoming), but lack of space prevents me from developing the argument here.

17. In the Rio Alamar in Tijuana, one of many runoffs for maquila effluents, five times the permitted level of mercury, as well as other toxics, were discovered. In the words of the article: maquilas 'consume and produce vast quantities of unsafe materials. Solvents, acids, resins, paints, plastics, oils, varnishes, heavy metals and pesticides are among toxins left over after production.' This piece has a useful map and checklist of maquila-induced trouble spots along the border.

18. The plant in question, Quimica Fluor in Matamoros (two-thirds owned by the Mexican Compania Minera Frisco, one-third by Du Pont), was so dangerous that it was the subject of Mexico's first 'Intermediate Safeguard Zone'. This meant that no new building was permitted within 1.25 miles of the plant. Locals were quoted as saying: 'move the plant, don't stop us living here', while experts advised that a five-mile zone was needed. Fluor claims to have invested $15 million in safety equipment. The *Los Angeles Times* notes that a similar plant in Texas is not sited in a residential area.

19. A Chicago-based waste management firm (see Sklair, forthcoming, note 11) moved into Tijuana in 1988, but its incinerator licence was revoked in 1992 when it was found to be about to burn millions of gallons of PCB oils, a known carcinogen. It now processes 800 fifty-five-gallon drums of toxics a month and ships them to the United States for burial. This represents 10–25 per cent of what is actually produced, and the manager of the firm is quoted as saying: 'It would be really interesting to know where the rest goes' (Ross, 1992, p. 5).

References

Barkin, D., 1990, *Distorted Development*, Boulder, Westview.

Barry, T. (ed.), 1992, *Mexico: A Country Guide*, Albuquerque, Inter-Hemispheric Education Resource Center.

Bath, C.R., 1992, 'The emerging environmental crisis along the United States-Mexico border' in L. Herzog (ed.), *Changing Boundaries in the Americas*, San Diego, University of California, Center for US–Mexican Studies.

Corona, R., 1991, 'Principales Caracteristicas Demograficas de la Zona Fronteriza del Norte de Mexico', *Frontera Norte*, **5**, January–June, pp. 141–56.

Day, M., 1992, *Stepan Chemical: the Poisoning of a Mexican Community*, San Antonio, Coalition for Justice in the Maquiladoras [videotape].

Farquharson, M., 1991, 'Is infrastructure holding up development in Tijuana?', *Maquiladora Newsletter*, **XVIII** (6), pp. 1–4.

Gassert, T., 1985, *Health Hazards in Electronics: A Handbook*, Hong Kong, Asia Monitor Resource Center.

Gilbert, A., (ed.), 1989, *Housing and Land in Urban Mexico*, San Diego, University of California, Center for US–Mexican Studies.

Guendelman, S. and Jasis, M., forthcoming, 'Health consequences of maquiladora work: women on the U.S.–Mexican border', *American Journal of Public Health*.

Hansen, N., 1981, *The Border Economy*, Austin, University of Texas Press.

Herrera Toledano, S., 1992, 'The ecological factor in the NAFTA', *Twin Plant News*, April, pp. 28–31.
Herzog, L., 1990, *Where North meets South: Cities, Space, and Politics on the U.S.–Mexico Border*, Austin, University of Texas, Center for Mexican American Studies.
Kelly, M., 1991, 'Facing reality: the need for fundamental changes in protecting the environment along the border', Austin, Texas Center for Policy Studies.
Kochan, L., 1989, *The Maquiladoras and Toxics: The Hidden Costs of Production South of the Border*, Washington, DC, AFL-CIO.
Leonard, H., 1988, *Pollution and the Struggle for the Global Product*, Cambridge, University Press.
Lorey, E., (ed.), 1990, *United States–Mexico Border Statistics Since 1900*, Los Angeles, University of California Latin America Center.
Lowery, S., 1990, 'Environmental services: a growing priority', *Twin Plant News*, May, pp. 45–8.
—, 1992, 'The border gets a voice', *Twin Plant News*, August, pp. 48–50.
Moure-Eraso, R., Wilcox, M., Punnett, L., Copeland, L. and Levenstein, C., 1991, *Back to the Future: Sweatshop Conditions on the Mexico–U.S. Border*, University of Lowell-Work Environment Program, mimeo.
Mumme, S., 1992, 'System maintenance and environmental reform in Mexico', *Latin American Perspectives*, **19** (1), pp. 123–43.
Mumme, S. and Nalven, J., 1988 'National perspectives on managing transboundary environmental hazards: the U.S.–Mexico border region', *Journal of Borderlands Studies*, **3**, pp. 39–68.
Ortiz Monasterio, F., 1987, *Tierra Profanada: Historia Ambiental de Mexico*, Mexico City, SEDUE.
Perry, D., Sanchez, R., Glaze, W. and Mazari, M., 1990, 'Binational management of hazardous waste: the maquiladora industry at the U.S.–Mexico border', *Environmental Management*, **14** (4), pp. 441–50.
Rey Romay, B. (ed.), 1992, *La Integracion Comercial de Mexico a Estados Unidos y Canada: Alternativa o Destino?*, Mexico City, Siglo Ventiuno Editores/UNAM, Third Edition.
Reygadas, L., 1992, *Un rostro moderno de la pobreza: problematica social de las trabajadoras de las maquiladoras de Chihuahua*, Chihuahua, Ediciones del Gobierno del Estado de Chihuahua.
Reynolds, C., Waverman, L and Bueno, G. (eds), 1991, *The Dynamics of North American Trade and Investment: Canada, Mexico and the United States*, Stanford, Stanford University Press.
Rich, J.G., 1992, 'Planning the border's future: the Mexican–U.S. integrated border environmental plan', Austin, University of Texas, US–Mexican Policy Studies Program.
Ross, J., 1992, 'Mexico's disappearing toxic wastes', *Latinamerica Press*, August, **20**, p. 5.
Sanchez, R., 1990, 'Otra manera de ver la maquiladora: riesgos en el medio ambiente y la salud' in B. Gonzalez-Arechiga and R. Ramirez, (eds), *Subcontratacion y Empresas Transnacionales*, Tijuana, COLEF, pp. 531–76.
—, 1990a, 'Condiciones de la vida de los trabajadores de la maquiladora en Tijuana y Nogales', *Frontera Norte*, **2**, July–December, pp. 153–81.
Sklair, L., 1991, *Sociology of the Global System*, Baltimore, Johns Hopkins University Press and London, Harvester.
—, 1992, 'The maquilas in Mexico: a global perspective', *Bulletin of Latin American Research*, **11**, pp. 91–107.
—, 1992a, 'The transnational capitalist class in Mexico and the USA: some evidence

from NAFTA', Los Angeles, LASA XVII International Congress.
—, 1993, *Assembling for Development: The Maquila Industry in Mexico and the United States*, San Diego, University of California, Center for US–Mexican Studies (new edition, enlarged and updated).
—, (forthcoming), 'Global sociology and global environmental change' in M. Redclift and T. Benton (eds), *Sociology and Global Environment Change*, London, Routledge.
United States Environmental Protection Agency/Secretaria de Desarrollo Urbano y Ecologia, 1992, *Integrated Environmental Plan for the Mexican–U.S. Border Area, (First Stage, 1992–1994)*, USEPA/SEDUE.
Waller, T., 1993, 'The politics of water conservation in the border region of the Californias', unpublished Ph.D. thesis, Sociology, San Diego University of California.
Wilk, D., 1992, 'The complementarity between environmental plans, regulations and economic incentives: lessons from Ciudad Juarez, Chihuahua', Los Angeles, LASA XVII International Congress.
World Bank, 1991, *World Development Report*, New York, Oxford University Press.

6 Low-income housing, the environment and mining on the Witwatersrand, South Africa

Malcolm Lupton and Tony Wolfson

The shift of South African society from a period of relative growth and stability to a deepening economic and political crisis, from the mid-1970s, has led to the emergence of a new body of urban research, informed by radical social theory. Significant as these developments and the rich body of literature they gave rise to may be (cf. Lupton, 1992), South African geographers, both human and physical, have yet to establish the close links between urban and environmental changes. A noticeable exception is the study of the environmental controls on low-income housing along the Natal coastal margins by McCarthy (1986). Moreover, the constraints of the physical environment on low-income housing provision in South African cities is an area which has hardly been explored, let alone explained in relation to the urban process. This is indeed surprising when it is considered that a major legacy of urbanisation under apartheid is the co-incidence of low-income housing, allocated according to racial criteria, with marginal and hazardous urban environments (cf. McCarthy, 1986). It is to fill this lacuna within the South African geographic literature that this chapter is directed.

The objective is to investigate the physical environmental constraints on low-income housing, specifically regarding the Witwatersrand, South Africa's main industrial and mining region. The study is perhaps not only instructive for cities and regions faced with similar problems but allows an analysis of the physical controls imposed on low-income housing by the imperatives of the gold-mining industry. It is moreover our contention, in contrast to the approach advanced by Brink *et al.* (1990), that the politico-economic determinants of the physical environmental hazards associated with Third World urbanisation be accorded conceptual primacy.

The chapter consists of three parts. First, the shift of South African society from a configuration of growth to stagnation and the concomitant transformations within the mining industry are examined as a context for discussing the physical environmental constraints imposed on low-income housing by the exigencies of an economy based on mining. Second, the mechanisms through which poor and powerless urban communities were relegated to medium- and high-risk land underlaid by dolomitic rock formations, or areas rendered marginal and hazardous by mining operations on the Witwatersrand, are investigated. In the last part of the chapter an analysis is made of the implications, for the urban poor, of reductions in state intervention and the attendant rise of a new market-orientated urban policy. Changes in South African urban policies will have profound social and spatial effects regarding low-income housing provision on the Witwatersrand. An important issue raised is whether low-income communities will be absorbed by the market or, alternatively, further relegated to environmentally risky and marginal urban areas. Certain policy-relevant proposals, which flow from the analysis, are presented in the concluding remarks.

The politico-economic crisis of apartheid and the transformation of South African gold mining

South Africa experienced unprecedented economic expansion and political stability after the Second World War. Between 1962 and 1966 South Africa's annual real growth rate of nearly 7 per cent was the second highest in the world, surpassed only by the 9 per cent of Japan (*Financial Mail*, Johannesburg, 6 September 1968). Although the mining industry, more than any other sector, was crucial to this economic performance, the dominance of mining increasingly gave way, after the Second World War, to manufacturing (Wilson, 1972). Between the 1950s and mid-1970s South Africa experienced, alongside the 'bureaucratic authoritarian' regimes of Latin America (Brazil, Argentina and Chile), expansion of secondary industry, the concentration of ownership and control of capital and capitalist modernisation by a business/bureaucratic/military nexus (Cooper, 1991).

Moreover, the South African economy became increasingly internationalised and exhibited features of monopolisation in all major sectors. The merging of mining and industrial capital and mining and bank capital was being initiated, giving rise to finance capital (Innes, 1984). Monopolisation was further accompanied by a growing role for the state in the economy. The extent of state involvement in South Africa has always been proportionally high, but during the 1960s the contribution to output by parastatals rose significantly (Innes, 1984). According to Innes, of ten concerns whose combined value of total assets in 1979 stood at R 50,268 million (US$ 1.00 = R 2.80, January 1992), four were state controlled (South African Railways and Harbours, the Electricity Supply Commission, the

Reserve Bank and the Iron and Steel Corporation). Significantly, the onset in the advanced capitalist countries of a global economic crisis, sparked by the oil price shocks of 1973/4, had the same effect in all these older-generation NICs.

From the late 1970s, the authoritarian regimes which had presided over the spectacular paths of industrialisation in Latin America and South Africa descended into a deepening economic and political crisis. The growth model of these older NICs tied them to the declining fortunes of the advanced capitalist countries of Western Europe and North America, and the introduction of Western growth strategies, geared toward externally orientated industrialisation, was not accompanied by the development of internal markets of mass consumption (cf. Lipietz, 1987). Paradoxically, it is the monetarist policies adopted in response to the crisis in the advanced countries, combined with internal factors within the NICs, that destroyed many of the miracle economies of the 1960s and 1970s (Lipietz, 1987), including South Africa's.

An important consequence of the stagnation in the manufacturing sector is that mining has once more assumed a prominent position in the South African economy. In the mid-1980s gold returned to represent more than half of the value of South African goods exported and provided approximately 10 per cent of the state's budget in direct taxes (Freund, 1990). Indeed, a rising proportion of the profits of the Anglo American conglomerate was derived from mining, while industry accounted for under 20 per cent (*Financial Mail*, Johannesburg, October 1987).

Further stagnation in manufacturing is likely to lead to the export of more primary commodities like minerals. Expenditure on R & D and exploration rose substantially after the late 1970s as the mining industry sought to extend existing mines, re-exploit abandoned mines with new technology and probe exceptionally deep levels (Freund, 1990). Within a context of politico-economic crisis, increasing reliance by the South African economy on an environmentally destructive industry, such as mining, raises important questions pertaining to low-income housing provision in areas, such as the Witwatersrand, whose entire urban system has historically been organised around the mining industry.

Physical constraints on low-income housing in a mining-based economy: the case of the Witwatersrand

The Witwatersrand, engine of the South African economy and site of some of the world's richest gold deposits, is a region based on mining, arguably the most hazardous and environmentally destructive industry. Covering approximately 2 per cent of the national space, this region is the most urbanised part of South Africa; it accommodated more than 23 per cent of the population and produced 43 per cent of the country's GNP in 1985 (Beavon, 1989). The estimated population for 1992 is 7.3 million and is expected to grow to twelve million by the year 2000 (Beavon, 1989). Opening of the gold mines of the Witwatersrand a century ago represents the first modern large-scale industrial activity in South Africa. The Witwatersrand is also of great significance because it is here that a familiar pattern of South African urban development first appeared: the co-incidence of low-income, and mostly black, urban communities with hazardous and marginal physical environments such as was produced by the forced relocation of blacks in 1904 to a municipal sewage dump (Beavon, 1982). The marginalisation of the urban poor to peripheral locations, often at great distances from employment centres and services, is closely connected with the very nature of gold mining on the Witwatersrand.

Mining activity and the marginalisation of the urban poor

This marginalisation of the urban poor by mining activity on the Witwatersrand has come about in various ways. First there are the direct physical controls imposed by the mining industry over the spatial distribution of low-income housing. The geological structure of the Witwatersrand proved a crucial determinant of gold mining in the region (Johnstone, 1976). Unlike the diamond deposits at Kimberley which were easily available on the surface to individual claim holders, the gold-bearing formations of the Witwatersrand sloped deep underground, from east to west. The average grade of gold in the ore was very low which meant that extraction necessitated capital-intensive operations involving deep-level mining and large and complex reduction works, especially on the western Witwatersrand. In contrast to open-cast and strip mining, deep-level extraction of gold required high capital outlays (Johnstone, 1976). On the Witwatersrand, the costs incurred by deep-level mining were compounded by the abrasiveness of the rock formations overlying the gold-bearing ore. Moreover, up to the oil price shock of 1973/4, the price of gold was set at international rates which, combined with low-grade ore and high overhead capital costs, produced a specific costs structure for the Witwatersrand mines. Profits could be secured only through a very low level of costs and a high level of output (Johnstone, 1976).

Cheap black labour was a solution since it enabled the mining companies to reduce the price of labour, a key determinant of production costs. The function of racial policies in southern Africa, in securing a steady supply of cheap black labour for the mines, and later secondary industry, through various mechanisms such as the alienation of indigenous African land by the Land Acts of 1913 and 1936 and the migrant-labour recruiting system, is well documented (Johnstone, 1982). Of significance here is that the emerging spatial distribution of those drawn into wage labour on the Witwatersrand coincided with hazardous and marginal urban environments. The profitability of gold mining on the Witwatersrand, and hence the performance of the South African economy, ultimately depended on a secure supply of cheap black labour (Johnstone, 1976).

To maintain low labour costs, a reduction of the level of the average wage rate in all economic sectors became imperative, as demonstrated by the importation of cheap Chinese labour as a means to reduce aggregate demand for African labour throughout British South Africa, thereby lowering the average wage rate (Richardson, 1977). One strategy, specific to the mining industry, was to establish a centralised labour recruiting system (Crush *et al.*, 1991) to minimise competition for labour among mining companies. Another strategy, operating for the entire economy, was the location of housing for the workforce on marginal inexpensive urban land. A major determinant of the level of the wage rate, the costs of accommodation for the workforce, was thus reduced.

Historically, low-income accommodation on the Witwatersrand was situated on cheap urban land; areas residual to mined land. The allocation of land proclaimed for mining purposes by the state to mining companies also withdrew large portions of suitable, accessible land from residential use. The ownership by mines of large tracts of land on the Witwatersrand (Figure 6.1), some within walking distance of the CBD of Johannesburg, has greatly influenced the location of low-income housing on marginal physical locations. Peripheralisation of poor and unemployed communities on the Witwatersrand has occurred largely as a result of the powerful control the mining companies have exercised over the allocation and use of urban land in the region.

Second, mining operations along an east–west axis on the reef directly controlled land use patterns, the location of the CBD of Johannesburg and infrastructural development such as transportation, tertiary institutions and health services. Physical constraints, particularly south of the proclaimed mining land, combined with suitable residential land in the north, encouraged the emergence of different land markets on either side of the mining belt which dissects the city (Figure 6.1). Whereas the railway line previously segmented Johannesburg into a north/south spatial divide, the series of ridges which pass through the city, from east to west, formed the new divide. As Johannesburg expanded, the affluent, like the first mining magnates, whose property values were deflated by the encroaching working class, relocated over the ridge to the north of the city into salubrious suburbs like Parktown (van Onselen, 1982).

Access to substantial financial resources enabled the *nouveau riche* to invest in property in the northern suburbs, thus inflating the value of space on what was previously cheap agricultural land. During the inter-war period, low-density, upper-income housing spread into areas in the

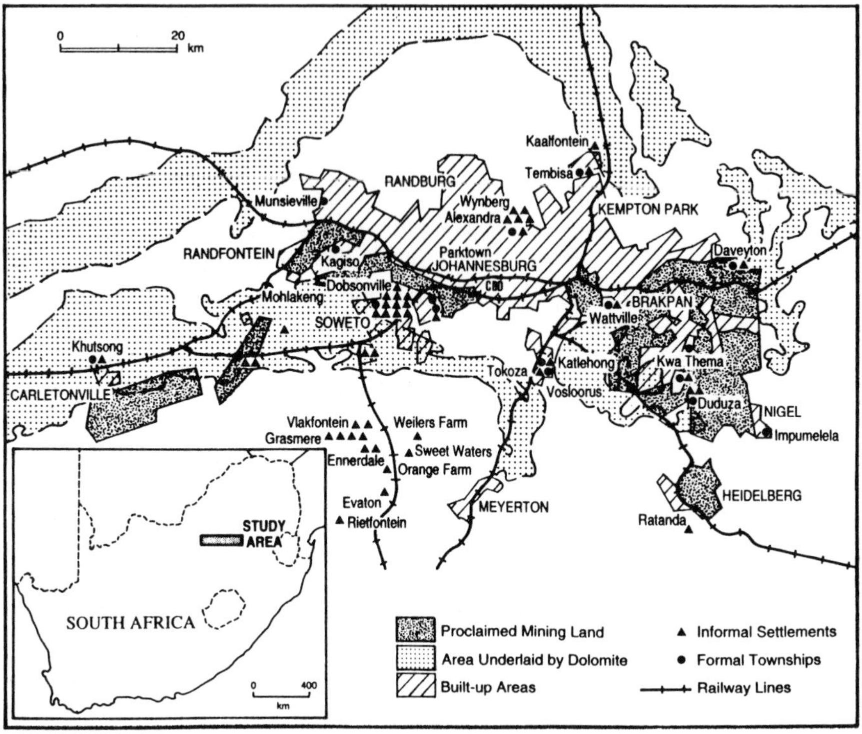

Figure 6.1 The environmental constraints to low-income housing on the Witwatersrand

north of Johannesburg, while higher-density housing, often with middle- and low-income residents, spread to the south, east and west, adjacent to mining and industrial activities and the CBD. Prior to the implementation of apartheid policies in the 1950s, Johannesburg had a loose, zonal pattern, with residential segregation brought about partly by informal social pressures, partly by choice and partly by a range of local by-laws and title deed restrictions (Infraplan, 1992). The Group Areas Act, passed in 1950, introduced a stringent urban land-use planning framework with distinct features such as discrete residential areas for each racial category, rigidly separated by buffer strips of vacant land or physical boundaries such as railway lines, light industries and natural barriers.

The location of these racially segregated locales met the politico-ideological requirements of apartheid urban planners and, additionally, they were not so distant as to make the provision of transport subsidies for the workforce unviable for the state. The location of low-income housing on the urban periphery was shaped by the operation of the urban land market, politically motivated racial segregation and juridical controls which pertain to those regulations governing the use of proclaimed mining land for purposes other than mining. Only recently were legal provisions made for including proclaimed mining land within the jurisdiction of the land use plans of local authorities. None the less, for the land use to be altered from mining to residential functions involves intricate legal procedures, and in most cases such land is unfit for human habitation as a result of contamination by past mining activities.

In addition to the peripheralisation of the urban poor by political processes and the urban land market, mining operations further imposed geophysical constraints on future land uses. The sinking of shafts, sometimes as deep underground as 3.7 kilometres, causes slope instability. The deepest mine shaft in the world is at Western Deep Levels on the far west of the Witwatersrand, with a depth of 3,793 metres, where mine workers deal with a rock surface temperature of 55°C (Elliot, 1986). Moreover, pumping of water from vast subterranean aquifers in the dolomitic rock formations which overlie the gold-bearing ore (also referred to as dewatering) is required (Brink *et al.*, 1990). Dewatering is an expensive operation which depletes a natural water source and leads to subsidence of the water-table.

Dewatering causes the subsurface water content to change, thereby affecting the stasis of the dolomite. The re-designation of land mined at depths of less than 240 metres, for low-income housing provision, is precluded as a result of excavation which has rendered the subsurface rock structurally unstable. Subsidence can occur on land which has been undermined by deep-level excavations, and the results can be devastating for people living on such land, especially vulnerable low-income communities. Once the stasis of dolomite is disturbed, the result is often a collapse of cavities within the dolomite rock formations, causing sinkholes or dolines (Figure 6.2).

A major environmental hazard relates to mine dumps and slimes dams, both waste products of mining. The solid wastes produced by the reduction process through which gold is separated from the ore are deposited in dumps, while liquid waste products are pumped into slimes dams. There is a varying uranium content which is extracted with the gold-bearing ore. Uranium is a radioactive substance which, on decaying, emits an inert gas called radon. This gas escapes more easily from finer mine dump particles than it would from underground, where it occurs naturally. Although not dangerous in small quantities, the quantity of radon emitted by radioactive material in mine dumps and slimes dams poses a definite health threat:

> When inhaled, and while in the lungs, radon may decay further and become a solid radio-active particle which attaches itself to the lung tissue. With the formation of each subsequent radio-active . . . product, the lungs continue to be irradiated for at least another 22 years. [City of Johannesburg, 1992, p. 1].

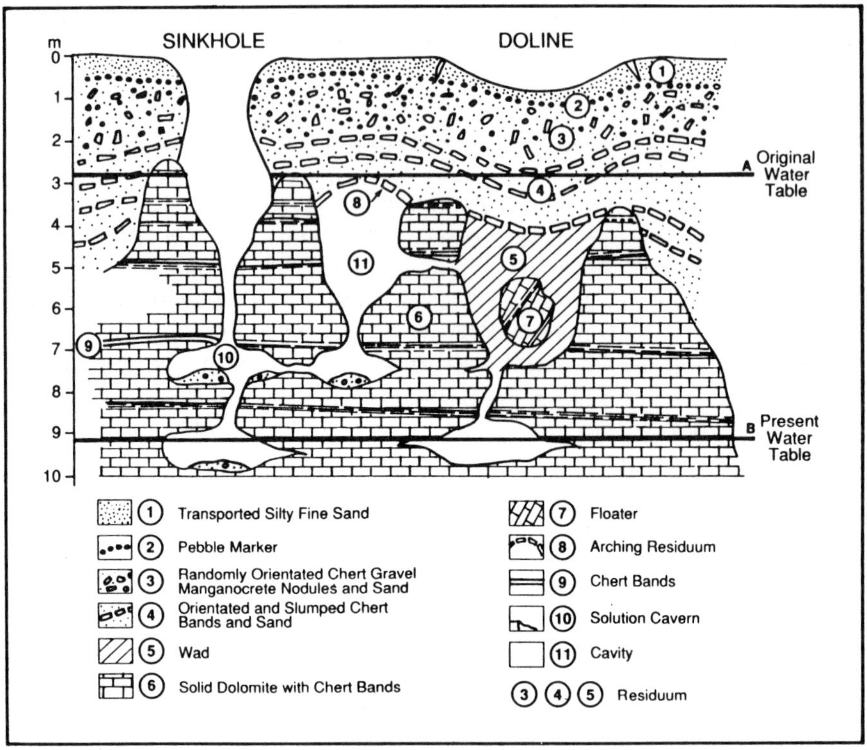

Figure 6.2 Mechanism of doline and sinkhole formation (adapted from von Wagener, 1985)

Furthermore, respiratory diseases such as silicosis are directly related to dust blowing from mine dumps (City of Johannesburg, 1992).

Evidence also suggests that additional health hazards are caused by the acids and chemicals used in the reduction process which contaminate streams flowing through areas of mining activity. For example, a stream flowing along mine dumps and slimes dams, from Crown Mines to Soweto, has revealed 'exceptionally high levels' of uranium, sulphate, cyanide and arsenic (Coetzee and Cooper, 1991, p. 134). A laboratory analysis revealed that the water contained eight million micrograms of sulphate, 1,900 micrograms of uranium, 520 micrograms of cyanide and sixty micrograms of arsenic per litre, far exceeding the standards for potable water established by the World Health Organisation (Coetzee and Cooper, 1991). On the Witwatersrand, there is an abundance of land contaminated and made 'sterile' by mining operations. Given the current economic crisis and the costs associated with rehabilitation, allocation of such land for low-income residential purposes is precluded. Also, the engineering difficulties related to the provision of urban infrastructure and services on land with mine dumps and inadequate drainage effectively render mining land unsound for human habitation.

The coincidence of low-income housing with the dolomitic belt

An unintended consequence of the marginalisation of the poor to hazardous and peripheral locations is the coincidence of low-income housing with land underlaid by dolomitic rock formations. In the Johannesburg metropolitan region alone, most of these low-income settlements, located on the urban periphery, are underlaid by soluble dolomitic rock formations which surround Johannesburg, the hub of the Witwatersrand region (Figure 6.1). It is estimated that more than 14 per cent of the Witwatersrand is underlaid by near surface dolomite (van Schalkwyk, 1981), which also poses a greater risk to low-income accommodation because of seepage from unserviced sewage and drainage.

Although dolomite varies in composition, there are nevertheless certain prevailing geophysical characteristics (Figure 6.2). The surface is covered with a blanket of transported silty, fine sand, underlaid by a thin layer of pebbles, called the pebble marker. Below the pebble marker is the residuum, a layer of soil composed of gravel, sand and the wad. The water content of the residuum largely informs the stability of the dolomite. Cavities are often found in the residuum, between the pinnacles.

Most dolomitic rock formations are in a state of equilibrium or stasis; that is to say, the distribution of water and soil particles is stable. Mining and urban infrastructural development are the most common factors disturbing this stasis. For example, dewatering as a result of mining (Brink *et al.*, 1990), or the installation of underground service systems such as sewerage and water drainage pipes, causes a disruption of the stasis through water seepage. If underground water services are not managed and monitored, excess water seepage will erode the soluble components of dolomite. The pinnacles between the cavities can thus collapse, causing larger underground cavities, and hence the start of sinkhole and doline formation (Figure 6.2). Dolines occur with a slow subsidence of the ground surface, typically measuring fifty to 300 metres in length but sometimes stretching kilometres (Buttrick, 1990). Damage to surface structures can be substantial, but there is no danger to life.

A sinkhole, however, caused by the sudden collapse of surface material into a cavity ranging between three and fifty metres in diameter can be catastrophic. It can take up to twenty years for a sinkhole to develop which means that the environmental problems caused by dewatering as a result of deep-level mining or insufficient water management by mining companies and local authorities (monitoring of sewerage, drainage and other water-related services) will be felt for generations. The most common triggering mechanism of sinkhole formation is the seepage of water, by artificial means, into the soils overlying the dolomite bedrock (Pape, 1990).

Damage to the built environment and loss of life arising from geological problems have been far more severe on dolomite than on any other geological formation in South Africa. Residential areas for the urban poor in South Africa are characterised by low-cost housing units and high population densities. Most locales accommodating low-income populations are completely underlaid by dolomite (Katlehong near Germiston, Khutsong near Carletonville, Tokoza near Alberton and Vosloorus near Boksburg), whereas others such as Tembisa, near Kempton Park, and Soweto are partly underlaid by dolomite (Table 6.1). On average, five to six sinkholes are reported per month during the summer months on the Witwatersrand (Buttrick, 1990).

A village for mineworkers, 120 hectares in extent and comprising thousands of low-income housing units, was built on the western Witwatersrand (von Wagener, 1985). A geotechnical investigation was carried out, and when it was discovered that the settlement was underlaid by dolomite, high- and low-risk zones were identified for different types of development: low risk for housing and high risk for sportsfields. Despite these precautions, twelve sinkholes, associated with poor drainage during or after construction and water seepage, occurred in the area in five years. In one of the more notorious cases, Khutsong, a low-income formal housing settlement in the western Witwatersrand (Figure 6.1), twenty-four sinkholes have occurred over ten years (Pape, 1990).

Concern for the safety of the public and buildings led to the enactment

Table 6.1 Low-income settlements on the Witwatersrand underlaid by dolomite

	Approximate area underlaid by dolomite	
Residential area	in hectares	as percentage of residential area
Katlehong	1,100	100
Khutsong	170	100
Tembisa	560	35
Thokoza	200	100
Vosloorus	900	100

Note: The above figures are rough estimates drawn from 1:250,000 geological maps of the Witwatersrand and recent street maps together with 1:50,000 topographical maps of the same area

Source: Pape, 1990

of a provincial ordinance in which certain regulations pertaining to township proclamation were established (Province of the Transvaal, 1965). In terms of this ordinance, it became necessary for a township developer to report, among other things, on the suitability of the site with regard to soil and the presence of dolomitic rock. A further step was taken when the Town-Planning and Townships Ordinance of 1986, which amended the 1965 ordinance, made geotechnical investigations compulsory for areas underlaid by dolomitic rock (Province of the Transvaal, 1986). Most residential areas housing low-income communities on the Witwatersrand which are located on dolomite were established prior to 1965: Katlehong was established in 1945, Tembisa in 1957 and Vosloorus in 1963. Significantly, all these townships originated out of forced removals of low-income communities to marginal physical environments.

Although urban property developers are legally obliged in South Africa to conduct geotechnical investigations on land underlaid by dolomite, such investigations are expensive and thus tend to escalate the price of a stand, among other development costs. It is therefore not profitable for property developers to target land which requires expensive pre-construction engineering costs for urban residential development. Consequently, there is not much demand, on the part of private developers, for large tracts of land located on the dolomitic belt which underlies the urban periphery. In South Africa, the private housing market caters only for 20 per cent of the urban population (de Vos, 1987) since the provision of accommodation to the poor by financial institutions and private developers is not profitable.

From the mid-1980s, the state has therefore responded to the accommodation needs of the poor by allocating large tracts of cheap land to informal settlements. Unlike the Latin American experience, for example, land invasions which culminate in the proliferation of shanty towns are isolated exceptions to the rule. In South Africa, the term 'informal settlement' designates unserviced shanty towns composed of rudimentary shacks in contradistinction to the fully serviced formal housing units constructed and managed either by the state or by private developers for renting or selling. According to official policy regarding the Witwatersrand, as elaborated in 1989, these settlements were to be joined with existing townships to form a continuous belt of low-income housing stretching from Katlehong, Tokoza and Vosloorus (Katorus), in the central Witwatersrand, to Kwa Thema, Tsakane and Duduza (Kwatsaduza) in the south-east (Figure 6.1) (Mavuso, 1989). The state has initiated the purchase of land between these six townships, some of which are underlaid by dolomite. This proliferation of informal settlements on peripheral land underlaid by dolomite poses a major physical environmental problem. What is at issue here are the hazards and threats posed by urban policies for low-income communities, to which attention now turns.

Urban environmental location controls and the implications for low-income housing on the Witwatersrand

As the provision of low-income housing is subjected to the supposed 'rationality' of market forces in South Africa during the 1990s, important questions arise. Will this lead to an absorption of the urban poor and powerless by the market into environmentally sound locations? Alternatively, will the urban poor and powerless be further relegated to hazardous and marginal physical locations such as the dolomite girdle which surrounds the Johannesburg metropolitan region? To assess the implications of the new urban policy for low-income housing in South Africa, it is necessary to confront, in detail, the ideological parameters of the new policy, the financial mechanisms envisaged and the possible decision-making system that might emerge in implementing it. The new policy proceeds from the assumption that racial controls over the allocation of scarce resources such as urban goods and services have become dysfunctional to the economic order (Swilling, 1990).

Taken together, this policy framework consists of a clear deracialised ideological vision for the future of South African cities; up to R 3,000 million in loans to be targeted at home buyers; a R 2,000 million special 'development' trust fund underwritten by the treasury; and a new, market-driven system of municipal funding for services (Swilling, 1990). What is important is the unconscious logic, and hence the effect, these market-orientated policy initiatives of big business and government will have on the provision of low-income housing on the Witwatersrand in particular and on South Africa at large. As pointed out by critics (Bond, 1990; Lupton, 1993; Morris and Hindson, 1992; Swilling, 1990), the two crucial issues at stake are the *social* and the *spatial* implications of neo-liberal urban policies. Moreover, there are environmental consequences for low-income housing provision on the Witwatersrand.

Deregulation, privatisation and promotion of the 'free market', all under the guise of removing racial controls over the allocation of urban facilities, will commodify land and urban space according to market-based criteria, irrespective of the social and spatial effects. As Swilling (1990, p. 30) puts it, 'when poor people live on land and in houses that have become more "valuable" in market terms than what they can "afford", they will slowly but surely get pushed out by the relentless pressure of better-off owner-occupiers and by various forms of investment capital that require the properties for a range of purposes (speculation, CBD development, urban development, etc.)'. The decisive variables in the equation which will determine whether low-income communities on the Witwatersrand will be absorbed by the market, or further relegated to marginal urban environments, are: the impact of the housing market on urban land use and price; the political process; effective demand; and unsuitable geological conditions for housing (Wolfson, 1991).

Regarding land use patterns on the Witwatersrand, competition for the use of land is less severe on the periphery of the Johannesburg metropolitan region. Several reasons could be put forward to explain this: first, the neo-classical location theory of the greater the distance, the higher the cost of transport, the cheaper the land to compensate for other costs; second, the high costs incurred in developing land underlaid by dolomite; third, the historical evolution of spatial and social segregation in the Johannesburg metropolitan region as low-income communities were historically located in the south while the more affluent lived in the north. In the south, dust from mine dumps and their lack of aesthetic appeal are a negative environmental externality for upper-income housing. As Hart and Browett (1976) have shown, the price of urban land is in part dependent upon transport costs as well as distance from blighted and polluted mining and industrial areas. Given these constraints, it is unlikely, within this context, that less hazardous physical environments can be allocated to low-income communities.

Without housing subsidies or state intervention in the urban land market, the poor will be unable to enter this market and hence gain access to cheaper land and urban services. Even if land is affordable, low-wage earners would find it difficult to develop their sites unless local government town planning regulations were substantially changed to accommodate the deregulation of urban housing standards. As Pape (1990) points out, however, building standards for low-income housing located on dolomitic rock formations should not be relaxed. Current urban policy, which is directed at the deregulation of housing provision, entails the relaxation of safety and building standards for low-income areas. The Less Formal Townships Establishment Act of 1991 formalises the tendency towards deregulation and the growth of informal settlements, thus effectively absolving business and government of any political responsibility for providing safe, affordable housing.

At the political level, local racial and class interests and prejudices could also directly influence the willingness of local authorities on the Witwatersrand to allocate suitably located land for low-income housing. A recent example is the pressure brought to bear on provincial authorities by the Greenbelt Action Group (GAG), a wealthy lobby group, to halt the development of any informal housing in an area 'where greenspace has to be preserved' (*Star*, Johannesburg, 11 December 1991). What is probably at issue here is the 'not in my backyard', syndrome: that is to say, the 'preservation' of high-income areas from encroachment by negative externalities, in this case poor communities. The effect of the political resistance of high-income communities, such as the GAG, will be to relegate low-income communities to marginal urban environments characterised by hazardous conditions. The consequences are also clear in terms of the significance of political power in gaining access to prime urban environments. There is a distinct possibility that politically powerless communities will increasingly be relegated to marginal urban environments precisely because they do not have access to decision-making processes owing to a lack of organisation. In addition to participation in decision-making, access to a suitable delivery system is dependent on a viable and cohesive grass-roots social movement, which enables such organised communities to obtain funds from bodies such as the Independent Development Trust or the Regional Housing Boards. The reality, however, is that for the vast majority of South Africans the effective demand for necessary urban goods and services does not exist.

Approximately 80 per cent of South Africa's population cannot afford even a low-cost formal housing unit of R 26,000 at current prices, including the land (de Vos, 1987). On the one hand, the trend to deregulation and high-population-density, low-cost housing suggests that the region will be dominated by small (240 square metres) sites within informal settlements. Low-density housing on the other hand implies the perpetuation of underutilised prime urban space by the middle- and upper-income groups. For example, in Bryanston an approximate building density of 3.5 units per hectare prevails, whereas 10 kilometres away in Alexandra the density is 160 units per hectare (Wolfson, 1991). As Buttrick (1990) has shown, there are some low-income areas which have 2,000 per cent more persons per square metre than wealthy suburbs. Given that South African cities do not display high building densities but high population densities, population growth and the underutilisation of prime urban space will result in the proliferation of informal settlements and continued urban sprawl. Approximately 2.26 million people, that is to say half of the black population of the Witwatersrand, already live in informal settlements (Beavon, 1989). While the physical environmental constraints and hazards associated with mining operations and land underlaid by dolomite persist, as shown above, research indicates (Beavon, 1989) that the most vulnerable sectors of the population, those who reside in the informal settlements on the Witwatersrand, are rapidly growing.

Conclusion

The argument advanced in this chapter is that the marginalisation of low-income and vulnerable urban communities on the Witwatersrand is directly related to two factors. First, there are the environmental constraints caused by mining operations in the region. Radon contamination, dewatering, undermining, toxification of soil by acids and the dust from mine dumps have rendered vast tracts of urban land unfit for human occupation. Conversion of such polluted land for housing the urban poor is thus precluded. Second, the relegation of marginal and vulnerable communities to the urban periphery, which on the Witwatersrand is incidentally underlaid by dolomite, poses a major environmental hazard, especially where such rock formations have been rendered structurally unstable by deep-level mining. Low-income housing on areas underlaid by dolomite is exposed to the risks of sinkholes and dolines caused by mining as well as by seepage from unserviced sewage and drainage.

The shift in South African urban policy, from excessive state intervention along apartheid lines to subjecting the provision of affordable housing to market forces, raises questions pertaining to the future spatial distribution of the urban poor and powerless. What has emerged is that low-income communities will be further relegated to marginal physical environments as vested interests, the mining industry among others, are allowed to dictate the social and spatial structuring of South African cities. The urban effect will be increased polarisation, conflict and endemic political violence as only a small proportion of urban residents gain access to urban goods and services while the vast majority are excluded (Morris and Hindson, 1992). For those, up to 80 per cent of South Africa's population, left out by the market, the only alternative is to enter the sprawling informal settlements on the urban periphery. The spatial effect will be a new urban form characterised by stark discrepancies between affluent, well-equipped and serviced middle- and upper-income suburbs, on the one hand, and poverty-stricken informal settlements on the other. In the case of the Witwatersrand, most, if not all, of the region's future affordable housing stock will be located on peripheral land underlaid by dolomite or on marginal environments, polluted by previous mining activity.

Several policy-relevant issues flow from our analysis. Contamination of soil and water by toxic chemicals and acids; the potentially harmful radiation emissions (radon) from slimes dams and mine dumps; dust blowing from the dumps, especially during the windy season: all these necessitate the rehabilitation of such land before it can be used for affordable or any other type of accommodation. Informal settlements should not be located on sites underlaid by dolomite or on land contaminated and toxified by the waste produced by mining operations. The implications are clear: safety and building regulations should not be deregulated or relaxed, and the distribution of the costs of rehabilitating toxified urban environments should be confronted. That is to say, who should pay for the destruction and contamination of the physical environment, and under what conditions? The mining industry? The state?

Given the engineering costs incurred in providing housing and services on dolomitic rock formations, and the constraints faced by informal settlers when seeking access to housing finance, it makes no economic sense to locate poor people on land which requires expensive geotechnical investigations for assessing risk and costly water management. Alternative sites for low-income communities should therefore be identified in areas which are environmentally safe as well as accessible to urban goods and services. Given the geophysical constraints of the Witwatersrand, the area immediately north of Johannesburg is the most environmentally suitable location for siting future high-density informal settlements.

Finally, as pointed out by the reconnaissance mission of the World Bank (1992) on South Africa's urban sector, there is a need for research that would identify which services and facilities are provided in which areas; to what extent they are efficiently utilised; and what potential such areas have for increased densification. This would directly relate to the problem pertaining to the efficient utilisation of existing infrastructure and services in all areas on the Witwatersrand. The evidence presented here shows that in areas of major mining activity, such as the Witwatersrand, a tension exists between the economic imperatives of the mining industry, on the one hand, and on the other the most basic safety and ecological needs, especially of low-income communities, such as stable land, a clean environment and wholesome water. This contradiction is aggravated as the South African economy relies more on an environmentally destructive industry which renders a growing proportion of potential residential land 'sterile'. Above all, the endemic violence which has engulfed the Witwatersrand is reflective of a highly divisive society emerging out of the ashes of the apartheid system as neo-liberal urban policies repolarise society along new class, ethnic and racial configurations (Morris and Hindson, 1992).

Acknowledgements

PLANACT is thanked for its resources and facilities made available, as well as for providing a stimulating environment for learning, without which this study would have been impossible. The contribution by Malcolm Lupton was funded by the Ford and Mellon Foundations. Mrs Wendy Job of the cartographic unit of the University of the Witwatersrand is thanked for the preparation of the diagrams. The ideas expressed, however, are those of the authors only.

Editor's Note

This chapter was written in 1992, before the transition to a post-apartheid state in the Republic of South Africa and related political processes picked up pace.

References

Beavon, K.S.O., 1982, 'Black townships in South Africa: *terra incognita* for urban geographers', *South African Geographical Journal*, **64**, pp. 3–20.

Beavon, K.S.O., 1989, 'Mexico City and *colonias populares*: hints for a South African squatter policy', *South African Geographical Journal*, **71**, pp. 142–56.

Bond, P., 1990, 'Township housing and South Africa's 'financial explosion': the theory and practice of finance capital in Alexandra', *Urban Forum*, **1** (2), pp. 39–67.

Brink, A.B., van Schalkwyk, A., Partridge, T., Midgley, D., Mall, J. and Geldenhuis, S., 1990, 'The changing impact of urbanization on the geological environment', *South African Journal of Science*, **86**, July–October, pp. 434–40.

Buttrick, D.B., 1990, *Characterisation and Appropriate Development of Sites on Dolomite. Perspectives of the unit for Applied Studies on Dolomite: Special report No. 1990-0054*, Johannesburg, *South African Geological Survey*.

City of Johannesburg, 1992, 'Radon: what is it? how can we deal with the problem?' Unpublished paper presented to a seminar on the effects of radon on land use in a local authority, Johannesburg, Health and Housing Department, 27–29 January 1992.

Coetzee, H. and Cooper, D., 1991, 'Wasting water: squandering a precious resource', in J. Cock and E. Koch (eds), *Going Green: People, Politics and the Environment in South Africa*, Cape Town, Oxford University Press, pp. 129–38.

Cooper, D., 1991, 'Locating South Africa in the third world: comparative perspectives on patterns of industrialisation and political trade unionism in South America', *Social Dynamics*, **17** (2), pp. 1–40.

Crush, J., Jeeves, A. and Yudelman, D., 1991, *South Africa's Labour Empire: a History of Black Migrancy to the Gold Mines*, Cape Town, David Philip.

de Vos, T., 1987, 'The complexity of housing in South Africa'. Unpublished paper presented at the 2nd Techno-Economics Symposium. Council for Scientific and Industrial Research, Pretoria, 28 January.

Elliot, H., 1986, 'Down towards the styx: an account of the development of Western Deep Levels', *Optima*, **34** (2), pp. 102–12.

Freund, W., 1990, 'South African gold mining in transformation' in S. Gelb, (ed.), *South Africa's Economic Crisis*, Cape Town, David Philip, pp. 110–28.

Hart, T. and Browett, J.G., 1976, 'A multi-variate spatial analysis of the socio-economic structure of Johannesburg, 1970', Occasional Paper No. 13, Johannesburg, Urban and Regional Research Unit, University of the Witwatersrand.

Infraplan, 1992, Johannesburg metropolitan residential viability study (phase II): commissioned by the Johannesburg Metropolitan Planning Department, Infraplan Town Planners (Pty.) Ltd., Johannesburg.

Innes, D., 1984, *Anglo American and the rise of Modern South Africa*, New York, Monthly Review Press.

Johnstone, F.A., 1976, *Class, Race and Gold: a Study of Class Relations and Racial Discrimination in South Africa*, London, Routledge.

—, 1982, '"Most painful to our hearts": South Africa through the eyes of the new school', *Canadian Journal of African Studies*, **16** (1), pp. 5–26.

Lipietz, A., 1987, *Mirages and Miracles: the Crisis of Global Fordism*, London, Verso.

Lupton, M.H., 1992, 'Economic crisis, deracialization and spatial restructuring: challenges for South African radical geography' in C.M. Rogerson and J.J. McCarthy, (eds), *South African Human Geography: Retrospect and Prospect*, Cape Town, Oxford University Press.

Lupton, M.H., 1993, 'Collective consumption and urban segregation in South Africa', *Antipode*, **25** (1), pp. 32–50.

Mavuso, J.S., 1989, Additional land for black settlement. Unpublished paper, conference on the Witwatersrand. Witwatersrand chamber of commerce and industry, Johannesburg, 27–28 June.

McCarthy, J., 1986, 'Problems of planning for urbanization and development in South Africa: the case of Natal's coastal margins', *Geoforum*, **17** (2), pp. 267–87.

Morris, M. and Hindson, D., 1992, 'South Africa: political violence, reform and reconstruction', *Review of African Political Economy*, **53**, pp. 43–59.

Pape, I., 1990, Development on dolomite in the Pretoria–Witwatersrand–Vereeniging area with specific reference to low-cost high-density mass housing schemes. Unpublished B.Sc. (Hons) dissertation, Johannesburg, Department of Civil Engineering, University of the Witwatersrand.

Province of the Transvaal, 1965, Town-planning and townships ordinance (No. 25), Pretoria, Government Printer.

—, 1986, Town-planning and townships ordinance (No. 15), Pretoria, Government Printer.

Richardson, P., 1977, 'The recruiting of Chinese indentured labour for the South African gold mines, 1903–1908', *Journal of African History*, **18** (1), pp. 85–108.

Swilling, M., 1990, 'Deracialised urbanisation: a critique of the new urban strategies and some policy alternatives from a democratic perspective', *Urban Forum*, **1** (2), pp. 15–38.

van Onselen, C., 1982, *Studies in the Social and Economic History of the Witwatersrand, 1886–1914*, vols 1 and 2, Johannesburg, Ravan Press.

van Schalkwyk, A., 1981, Patroon van ontwikkeling en risiko evaluering in dolomiet gebiede. Unpublished seminar paper: Engineering Geology of Dolomite Areas, University of Pretoria, Pretoria, 26–7 November.

von Wagener, M., 1985, 'Dolomites: problems of soils in South Africa', *The Civil Engineer in South Africa*, **27**, 395–407.

Wilson, F., 1972, *Labour in the South African Gold Mines*, Cambridge, Cambridge University Press.

Wolfson, T., 1991, 'Access to land' in M. Swilling, R. Humphries and K. Shubane (eds), *Apartheid City in Transition*, Cape Town, Oxford University Press, pp. 231–57.

World Bank, 1992, *South Africa Urban Sector Reconnaissance Mission: Aide Memoire*, Washington DC, World Bank.

7 The incidence and causes of slope failures in the *barrios* of Caracas, Venezuela

Virginia Jiménez-Díaz

Physical expansion during the past few decades has given rise in most Latin American cities to the occupation of the least stable locations within the cities, such as hillsides prone to landsliding and land subject to flooding.

One consequence of the alteration of natural environmental characteristics is that morphogenetic processes have become a risk to the communities (Cooke, 1984). As a result there are growing numbers of areas vulnerable to triggering factors such as earthquakes and heavy rainfall, among other so-called 'natural disasters'. The incidence of slope failures in the squatter settlements (*barrios*) of Caracas is such a case.

In searching for the factors that lead to slope instability in Caracas, physical characteristics of the environment as well as man-made changes as a consequence of urban growth need to be considered. This chapter examines both sides of the problem.

Urban growth, physical expansion and landslides in Caracas

Especially since about 1970, landslides in Caracas have been closely associated with the rainy season and with the creation of more vulnerable areas by the urbanisation process. Very few of the landslides can be considered natural. A common case is the establishment of urban areas that have been 'planned' without taking into account the suitability of terrain in terms of physical stability. Slope failures occur in *barrios*, not only because of the already unstable natural conditions but also as a result of human-induced modification to the slopes. Every year, increasing numbers of slope failures during the rainy season have endangered the lives and property of the inhabitants of these communities.

Barrios *and squatters*

In the case of Venezuela, squatter settlement presents a wide spectrum of housing conditions, ranging from precarious shelters to consolidated houses. Each shelter in the *barrio* is called a *'rancho'* and it is originally built of waste material or non-durable material. The author shares the view of Briceño-León (1986, p. 77) that 'not all the houses in the *barrio* are *ranchos*'. He believes that the definition should not be based on the quality of the houses, or the quality of the basic services (sewerage, water supply, electricity etc.). The criterion is 'the illegality or spontaneity or unpredictability of the way in which the land is occupied, and the singular way of construction by which it evolves into an established urban area' (ibid.).

When squatters invade an area they are viewed with displeasure, but the situation is often tolerated by the state. The complicity of the latter contributes to the production of *barrios* (Bolívar, 1989). As the political system is supported by popular votes, the settlement is later supported by the provision of the infrastructure and services necessary for the consolidation process (Camacho and Bolívar, 1987) as a populist way of getting support from low-income communities. In some cases upgrading programmes for *barrios* are limited to the aesthetic aspects of housing. The lack of continuity in the policies towards the *barrios* has contributed to the patchiness of their service supply. The investment and resources could have been more effectively allocated to social, legal and physical aspects of the problem. Physical conditions are considered only when landslides occur. As Davis (1987, p. 11) states:

> Any risk assessment of disaster prone cities reveals that the normal pattern of governments is to pay minimal attention to the issues of public safety in squatter settlements. The notable exception to this rule is obviously the period immediately after a disaster when the minds of politicians are temporarily focused on the issue. Reasons for the lack of normal concern include a number of factors: their 'illegality' which places them outside the sphere of many governments' concern; the limited resources of governments in poor countries which are focused on day to day survival needs, not on long-term planning; but perhaps the major reason concerns the formidable problems in introducing disaster preparedness or mitigation planning in the light of severe social, economic, political and technical constraints.

Urban growth of Caracas

Although Caracas has a history of more than four centuries, it is only in the past forty years that rapid population growth has become evident in terms of space. The ways in which the population has progressively spread into the valley, mainly eastwards from the city centre, and the configuration of the valley itself have played an important role in the occupation process. With increasing pressure on limited space, urban growth has taken place in the surrounding hills both in squatter settlements and in middle- and high-income developments. Caracas's urban development has been a result of different plans, some improvised, in response to different interests at different times.

By 1985, 61 per cent of the total population of the Metropolitan Area of Caracas lived in *barrios* (Briceño-León, 1986), and a substantial proportion of these were located on hillsides. Half or more of the land occupied by *barrios* belonged to the state (Jiménez-Díaz, 1992b, p. 83). Yet the physical stability of the areas occupied by squatter settlement has received little attention. Although some public work has been carried out to stabilise slopes, its purpose is usually corrective engineering, the cost of which is usually deemed unjustified for squatter settlements. In fact 67 per cent of the total area occupied by *barrios* in Caracas by the mid-1980s was unstable enough to justify the eviction of their residents. But riots like that of February 1989 have shown how difficult it would be to use such a justification in the *barrios* (ibid., p. 84).

Landslides in Caracas and understanding of the physical environment

Landslides were rare events in Caracas before the middle of the twentieth century: only twelve were recorded between 1800 and 1949, though others went unrecorded. During the 1950s and 1960s, they averaged about one per year, but this increased dramatically to an annual mean of about twenty-five between 1971 and 1979 and about thirty-three during 1980–7 (ibid., p. 86). A majority of slope failures reported in Caracas until the 1960s were associated with earthquakes, whereas from about 1970 slope failures began to be more noticeably associated with the incidence of rainfall. At the same time the growth of the city seems to have overlooked the physical condition of its environment.

The emergence of the earliest *barrios* in 1917 coincided with a document considered to be the first geotechnical map of Caracas (Singer and Feliziani, 1985). This anonymous map shows rock units, formations and terrains very accurately. It also recognised the existence of slope failures. In later years, the construction of roads, viaducts, aqueducts and reservoirs stimulated improved understanding of the geology of the areas being developed. But no attempt was made to link this knowledge with that needed to plan the city.

The earthquake of 1967 focused attention on the physical environment of the city. Geophysical studies were carried out especially in the eastern part of Caracas. In 1972 FUNVISIS (Fundación Venezolana de Investigaciones Sismológicas) was created by government decree, in an effort to co-ordinate research on seismology, antiseismic engineering and the geology of the active faults in Venezuela. The problems of slope failure were still largely ignored, even though the city experienced numerous emergencies in the 1970s as a consequence of geotechnical problems. Towards the end of that decade, documents of geotechnical zonation at detailed scales were produced in order to support the urban plans of some areas in Caracas.

An important contribution was made by the *Estudio Geotecnico del area Metropolitana de Caracas Sector Central* (Feliziani *et al.*, 1985), a document of scale 1:10,000 which aimed to give geotechnical information for central areas of the city.

Problems in the *barrios* were accentuated during the 1980s. Some of these neighbourhoods improved their infrastructural services, but even where drainage and sewerage and other services existed they tended to be of poor quality. Improvement of existing neighbourhoods is often very difficult owing to the combination of awkward terrain and its occupation by densely constructed buildings.

By the end of the last decade, both government and people had recognised the role of the physical environment and changes to it in increasing the vulnerability of the city. Multidisciplinary studies are now carried out to fill the gaps and the lack of information which for a long time dominated the topic – projects like FUNDACOMUN (*Fundación para el desarrollo de la comunidad y fomento municipal*), whose aim is to supplement and update information on *barrios* including characteristics of the physical environment. Different local groups, aware of the situation, have started to put pressure on appropriate government bodies to produce documents providing information from which decisions can be made in terms of relocation or improvement of areas at risk. Research institutions and universities are also making contributions towards better understanding of the problem and have produced valuable discussion documents. One of these (Baldó and Bolívar, 1989) argues the case against eviction on the grounds of slope instability.

Patterns of landslide occurrence in Caracas

In the course of the research on which this chapter is based (Jiménez-Díaz, 1992a), slope failures in the Metropolitan Area of Caracas were analysed both spatially and over time in order to elucidate some of the factors responsible for them. The main purpose of the research was to identify the relationship between *barrio* settlement and the incidence and severity of landslides in Caracas. A dataset of 205 slope failures which occurred over a six-year period (1974–9) was created. Besides location and date of occurrence, the dataset incorporated seasonality and cumulative rainfall values, lithology, slope angle, type of mass movement and location within the Metropolitan Area. This helped in identifying some of the necessary conditions for the slope failures to occur.

The main features of these slope failures were:

(i) Slope failures occurred mainly during the rainy season: 81 per cent of all 205 slope failures between 1974 and 1979 (see Figure 7.1);
(ii) The majority of failures (56 per cent) were in *barrios;* failures in other residential areas accounted for only 15 per cent, whereas 20 per cent were associated with road cuts;
(iii) More than 60 per cent of the failures were related to calcareous schist in the intensively folded Mercedes formation, where material is highly fractured and weathered;
(iv) Approximately 40 per cent of failures were associated with slope angles between 24 and 32 per cent (14°–18°), these angles being lower in the *barrios* during the rainy season than elsewhere in the city or at other times of year.

The main areas of landslide concentration since 1800 are shown in Figure 7.2. Most of these areas were in barrios (Figure 7.3), the exceptions

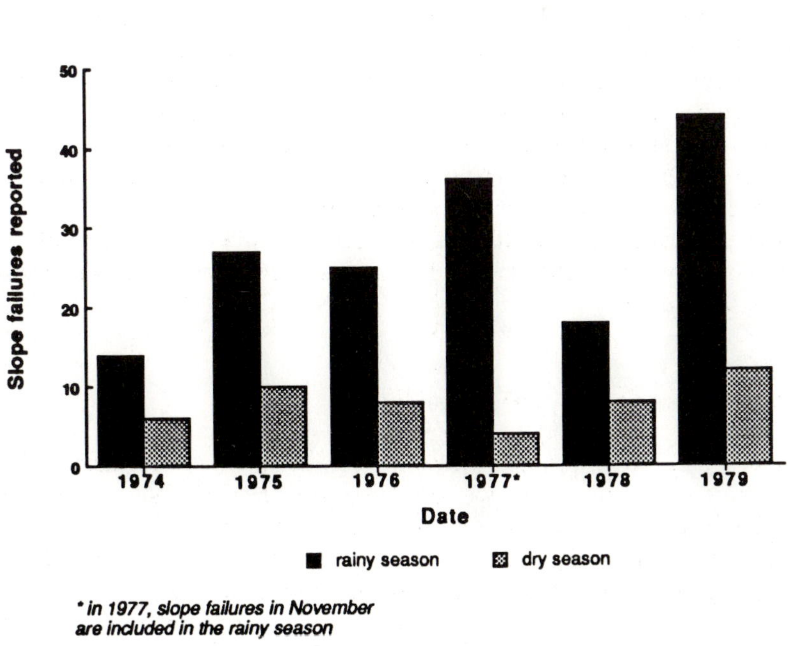

Figure 7.1 Seasonal distribution of slope failures in Caracas, 1974–9

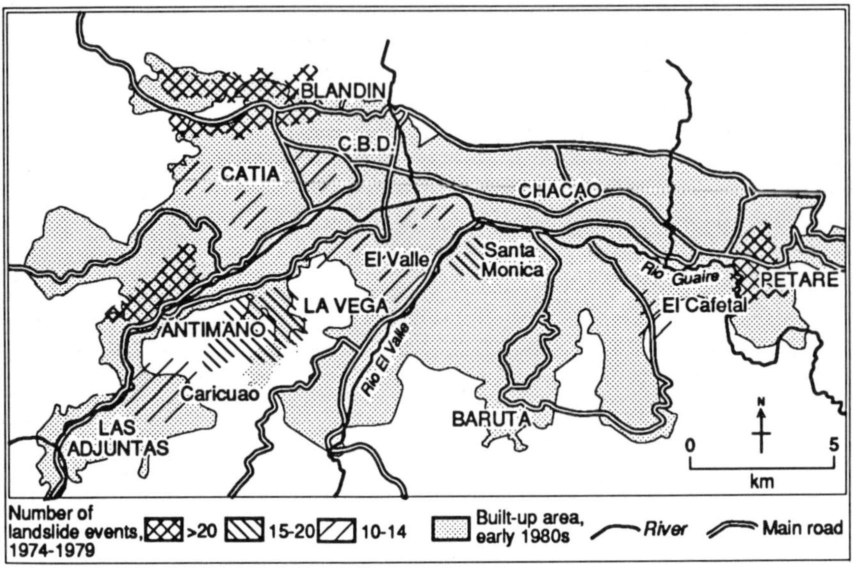

Figure 7.2 Main areas of landslide incidence in Caracas, 1974–79

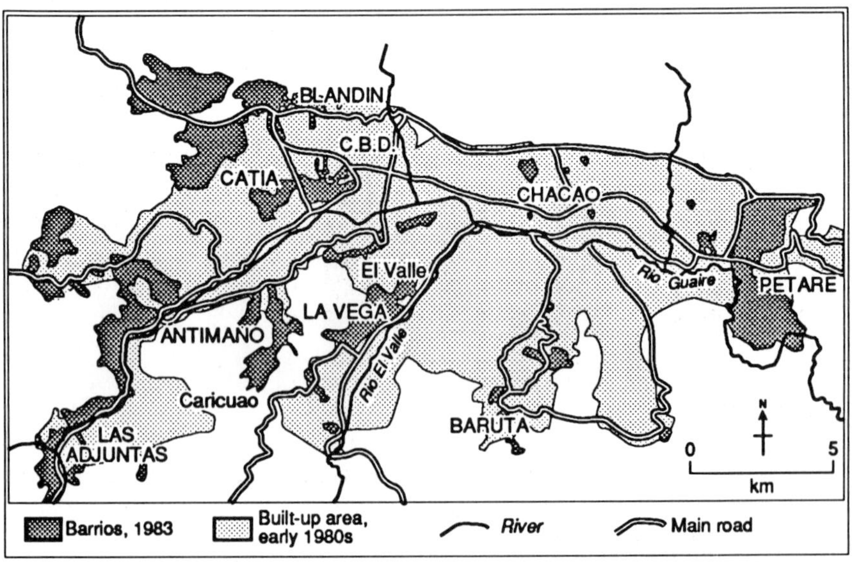

Figure 7.3 Areas of Caracas occupied by *barrios* in 1983

being Santa Monica and El Cafetal. The vulnerability of *barrios* to slope failure may be explained by a number of factors.

Physical conditions immediately antedating settlement in the *barrios* are important. Some of the areas studied did not present the most favourable conditions of physical environmental stability prior to settlement; one such area founded in 1957, *barrio* El Cipres, showed evidence of previous landslide activity in aerial photographs dating from 1953. Part of the unconsolidated material of the failure identified in the photograph was later occupied by the *barrio*, and the area again failed in 1983. No casualties were reported, but six houses were completely destroyed and a further twelve collapsed. Susceptibility to landslides increased where potentially unstable areas, with unconsolidated material present on slopes, were modified during the settlement process through steepening or loading.

Site clearance by cut and fill, in which soil is removed from the back of the site and deposited at the front, has been common in the construction of *barrios* on the hillsides of Caracas. This process provides more flat land for building, but the subsequent accumulation of unconsolidated material might also make such areas more prone to slope failure (Crozier, 1986).

Prior to *barrio* development, rainfall was intercepted by vegetation and distributed down natural drainage lines. With *barrio* construction, vegetation is cleared and replaced by closely spaced *ranchos* and paved areas, paths and stairs, which intercept rain and promote surface runoff; and this is supplemented by sewage and other waste water. Light rainfall might be disposed of without damage, but particularly heavy or continuous rain is now more likely to trigger landslide events, as runoff follows the new configuration of artificial channels in the *barrio* rather than the natural drainage lines. The fact that rainfall is a triggering factor can be of predictive significance, for it can warn of landslide hazard and of the need to activate local emergency services.

The replacement of building material when improving houses might be significant, but there is still not enough evidence to show how far the new, heavier materials used to improve the houses promote movement. How much weight a new unit can add harmlessly to the slope depends on the material and physical conditions characteristic of the slope, as well as on the care taken in site preparation and in building.

Conclusion

Analysis of the temporal and spatial pattern of slope failures has shown how physical environmental characteristics associated with man-made changes on the slopes might have increased the incidence of such events, particularly in *barrios*. In these areas, improvement in housing conditions may be the crucial element to consider in order to avoid major disasters.

More research is needed to answer the pressing issues raised here in order to reduce the risk of disasters within new squatter settlements of expanding cities, as well as in the crowded areas of settlements that already exist.

Editor's note

Landslides in Caracas on 8 August 1993, triggered by Tropical Storm Bret, killed over a hundred people and left thousands homeless in Petare, El Valle and other *barrios* indicated in this chapter.

References

Baldó, J. and Bolívar, T., 1989, *La investigación de los terrenos como causal de desalojo de los barrios caraqueños,* Caracas, SEU-FAU-UCV.

Bolívar, T., 1989, 'Los agentes sociales articulados a la producción de los barrios de ranchos', *Coloquio,* **1**, (1), pp. 143–61.

Briceño-León, R., 1986, *El futuro de las ciudades venezolanas,* Caracas, Cuadernos Lagoven.

Camacho, O. and Bolívar, T., 1987, *La propiedad y el inquilinato en los barrios de ranchos del Area Metropolitana de Caracas, una investigación necesaria,* paper presented at workshop on Land and Housing in Latin American Cities, University of California.

Cooke, R.U., 1984, *Geomorphological Hazards in Los Angeles,* The London Research Series in Geography 7, London, Allen and Unwin.

Crozier, M.J., 1986, *Landslides: Causes, Consequences and Environment,* London, Croom Helm.

Davis, I., 1987, 'Safe shelter within unsafe cities', *Open House International,* **12**, (3), pp. 5–15.

Feliziani, P., Barrientos, C., De Luca, P., Bastardo, S. and Ramirez, R., 1985, *Estudio geotécnico del Area Metropolitana de Caracas, sector central,* Caracas, VI Congreso Geológico Venezolano, Memorias, Tomo VII 4612-681.

FUNDACOMUN, 1978, *Inventario de los Barrios Pobres del Area Metroplitana de Caracas y el Departamento Vargas,* Caracas, Fondo Editorial Común.

Jiménez-Díaz, V., 1992a, *Slope failure in Caracas, Venezuela: the influence of squatter settlement,* unpublished PhD thesis, London, Department of Geography, University College.

—, 1992b, 'Landslides in the squatter settlements of Caracas; towards a better understanding of causative factors', *Environment and Urbanization,* **4**, (2), pp. 80–9.

Singer, A. and Feliziani, P., 1985, *Geología Urbana de Caracas,* Excursión N. 6, VI, Caracas, Congreso Geologico Venezolano.

8 Responses of low-income communities to environmental challenges in Hong Kong

Cecilia Chan

With almost six million people living on such a small piece of land, Hong Kong has the world's highest population density. Hong Kong produces 23,300 tonnes of solid waste, twenty-one tonnes of floating refuse and refuse from ocean-going vessels and dwelling boats, two million tonnes of sewage and industrial waste water every day, as well as 100,000 tonnes of chemical waste per year (Environmental Protection Department, 1990, pp. 21, 25). The Environmental Protection Department (EPD) has difficulty in identifying sufficient sites for landfills. At best floating refuse is a public eyesore, however passing untreated sewage into the harbour is potentially hazardous to public health. Typhoon shelters and bathing beaches are heavily contaminated. There are also serious problems of air pollution from industrial chimneys which are surrounded by newly built taller buildings. Noise from aircraft using Kai Tak Airport, situated in the midst of residential buildings, creates immense disturbance to the urban population. More generally, noise is a major nuisance that afflicts the population and may even present risks to mental health; indeed, residents in Hong Kong rank noise pollution on the top of a list of environmental problems (Lee, 1991). Noise emanates from a variety of sources including piling in construction sites, industrial/commercial operation, as well as motor vehicles and aircraft. Although the above environmental problems affect the general population in Hong Kong, it is the poor who suffer most from environmental degradation and pollution.

Despite the affluence in terms of Gross Domestic Product (US$12,069 per capita in 1990), there are large pockets of low-income housing with poor community facilities and environmental conditions in Hong Kong. The incidence of tuberculosis (6,704 new cases in 1989) and viral hepatitis (1,232 new cases in 1989) is still high, implying significant problems with nutrition and hygiene in the living environment. It needs little imagination to relate poor sanitation and environmental conditions to poverty.

Historical development

Historically, poor health, environmental hazards and infectious diseases have been associated closely with the poor living environment in Hong Kong. As early as 1841, when the British took over Hong Kong, the overcrowding, congestion, poor housing conditions, lack of proper sewerage and health care resulted in a high mortality rate especially among the Chinese population. The average age at death of a Chinese was only 18 years (Chan, 1991, p. 19). The first attempt by the government to address the public-health issue was the establishment of a Committee of Public Health and Cleanliness in 1843 to enforce the sanitary rules. In order to rectify the problems caused by inadequacy in building design and regulations, the government introduced a Building and Nuisance Ordinance in 1856. The appointment of a Sanitary Committee in 1862 followed the outbreak of cholera in the territory (Choi, 1993).

Osbert Chadwick was sent to Hong Kong by the Secretary of State in 1881 to examine the territory's conditions. In his report he claimed that the whole sanitary condition of Hong Kong was defective and recommended improvements in water supply, drainage, refuse and night soil collection, provision of additional public facilities plus a new building ordinance. However, the enforcement of his recommendations encountered great difficulties, especially from Chinese property owners (Chan, 1991, pp. 18–21).

In 1894, the years of neglect culminated in a bubonic plague which resulted in 2,552 deaths. This outbreak resulted in a number of ordinances and the establishment of a Housing Committee to look into dwelling conditions. The Closed House and Insanitary Dwellings Ordinance was introduced in 1894 to tighten construction controls, and the Tai Ping Shan Resumption Ordinance of that year authorised the government to redevelop the most afflicted areas. The health crisis in the 1900s gave impetus to a proposal for the physical segregation of the Chinese and European communities, which subsequently led to the passing of the Ordinance for the Reservation of a Residential Area in the Hill District in 1904. This ordinance, however, was repealed in 1946.

In the first three decades of the twentieth century, stringent dwelling controls and increased housing provision were prescribed as an answer to the territory's environmental problems. The possible contribution of town planning in this respect was not acknowledged until 1935 when the Housing Commission suggested the establishment of a permanent town planning and housing committee. In 1939, a Town Planning Ordinance was passed, and the Town Planning Office only came into existence after the Second World War in 1947 (Choi, 1993).

Environmental problems and hazards in Hong Kong

In general, natural environmental problems (such as rainstorm, typhoon, fire, landslides and flooding) affect the poor who live in squatter and temporary housing areas. However, man-made environmental hazards are degrading the quality of life of the majority of the population in Hong Kong (see Table 8.1). The most significant have been identified as LP-gas explosion; noise and gaseous emissions from factories resulting in high levels of particulates of sulphur dioxide and oxides of nitrogen in the atmosphere; lack of proper water treatment before discharge of waste water into the sea; noise and dust from construction sites; in the heavy traffic, exhaust smoke from diesel-engined vehicles; organic waste from

Table 8.1 Environmental problems in Hong Kong

Squatter settlements in Kowloon have been built on steep slopes where a lack of drainage and access roads results in high susceptibility to flooding, fires, sewage overspills and contamination, poor emergency access and other life-threatening problems. Factory locations often add to the poor air quality of low-income communities that tend to be built near to the sites of the heaviest polluting industries. The poor condition of the drainage system creates foul smells, allows mosquitoes to breed and leads to skin diseases, particularly among children, a large number of whom are reported to have coughs and respiratory problems, including chronic respiratory tract infections.

Specific environmental problems include the following:

- On the positive side general levels of sulphur dioxide recorded in 1991 have reduced dramatically (up to 80 per cent) in some industrial areas.
- Unleaded petrol was introduced in 1991, however, the majority of journeys use diesel-powered vehicles. Diesel engines contribute to about 80 per cent of nitrogen dioxide and nearly 98 per cent of traffic particulates. New regulations to control this pollution will not apply to buses, trucks and other large vehicles which create 70 to 75 per cent of this pollution.
- 90 per cent of sewage is untreated or only partially treated.
- There was a 51 per cent increase in solid waste between 1990 and 1991. At present the Government is subsidising the waste generators by providing free landfill space.
- 100 000 tonnes of chemical waste have been disposed of annually, mostly through the sewage system to the sea.
- The only construction work now under noise control involves the use of heavy machinery between 7 p.m. and 7 a.m. and on Sundays and other public holidays. Percussive piling is subjected to even stricter controls. However, other sources of construction noise are not restricted under present regulations.
- The Hong Kong government announced in 1991 Environment Impact Assessment (EIA) reports of public projects could be released for public consultation. However, access to these reports has been limited and no public hearing has been organised.

Sources: Douglass, 1992; Environment Protection Department, 1992; Ng, Cho-nam and Ng, Ting-Leung, 1992.

farming and domestic waste; dumping of chemical waste (Environmental Protection Department, 1991 p. 21).

The main source of pollution in Hong Kong is man-made industrial pollution. The conditions are succinctly described in the following terms by Hills and Barron (1990, p. 39): 'many streams are thick with industrial chemical wastes, residential areas are exposed to noxious factory smoke from short stacks, waterways carry untreated sewage, and industries often blatantly disregard existing environmental regulations'. It was only in 1989 that the government made a commitment to control these sources of pollution when it determined to take active steps in the area of environmental protection by establishing a White Paper on *Pollution in Hong Kong: A Time to Act*.

The impact of environmental problems in low-income communities

The poor are concentrated typically in three major locations: in temporary housing areas, squatter areas and urban slums. In the 1986 bi-census, there were 101,724 households living in temporary structures of squatter areas and temporary housing areas, 23,520 living in cockloft and bedspaces, 8,344 living in roof-top squatter huts and 162,554 living in room/cubicle accommodation or urban slums (Census and Statistics Department, 1990, p. 187). Thus, these 296,142 households, representing 20.48 per cent of households in Hong Kong, are living in sub-standard housing without self-contained facilities.

As with other aspects of their lives, the poor have little control over environmental problems. Frequently, these low-income communities are situated in the midst of industrial areas or by the side of highways. Consequently, residents suffer from the dual impact of pollution problems from adjacent industries and environment deterioration in the residential neighbourhood itself.

Squatter areas

These are the most flexible buffer to housing pressure in the metropolis, and the population of squatter areas expanded in line with the sudden influx of population to Hong Kong from mainland China in the 1950s, late 1970s and early 1980s. Indeed Triad Societies were involved in putting up squatter huts and selling them to new migrants who urgently needed to find affordable accommodation when they first arrived in Hong Kong.

Owing to the lack of coordinated planning, squatter huts are usually constructed along steep slopes without drainage systems and proper access roads. Water and electricity supply, refuse and waste-water disposal, hygiene and access are all problematic (Chan 1981; 1984), and in the mid-1970s for example, there were large-scale actions protesting against the lack of electricity supply to squatter areas. Clearly these problems pose a significant community health hazard. Thus the Council on Squatter Policy conducted a bacterial count of drinking water collected from squatter areas in 1981 and found that the quality of the drinking water was generally very poor. One water sample in Kwun Tong contained 500 times the tolerable standard of *E. coli* defined by the World Health Organization.

Temporary housing areas

In 1989, there were 98,300 inhabitants living in temporary housing areas managed by the Housing Authority in Hong Kong. Temporary housing areas are usually situated in remote suburban areas where the land price is still low. Low-income families who lose their homes due to squatter fire or demolition of private buildings for redevelopment and have lived in Hong Kong for less than seven years are eligible for placement in these temporary housing areas. Each person is allocated a floor space of twenty-four square feet, while kitchens are outside the living unit and toilet facilities are communal. The standard of hygiene of the toilets and baths is poor, and rats are common co-tenants in the highly congested temporary structures. As the temporary housing areas are, by definition, meant to be non-permanent, infrastructural provision is minimal. The sites are prepared by the Housing Authority with common toilets and baths, drains and refuse collection, space for small neighbourhood shops and community services as well as water, electricity and telephone supply. In terms of construction the huts are built with wood and tin sheets, and this poses a major problem in itself. Thus, in the summer they are very hot and the risk of fire is high. New problems are associated with residents who have lived in these temporary housing areas for over ten years and who find themselves living under precarious structures as termites eat away large parts of the pillars which support their dwellings. Furthermore, water drips into the huts during rainy days because of poor construction. Most of the temporary areas are located below flyovers, by the side of container depots or surrounded by industrial buildings; pollution caused by the noise, dust, and gaseous emissions from trucks and factories is severe.

Urban slums

Despite the Hong Kong government's efforts to construct public housing so as to provide decent accommodation for all, there are still 693,322 households living in private housing blocks. More than a quarter of these households share facilities with other households in the same living unit, often consisting of a bedspace, cockloft or small cubicle. Pre-war buildings without toilet facilities have to rely on night-soil collection. Urban decay and rapid deterioration of the infrastructure of pre-war and older buildings in the private market is causing problems of poor ventilation, lack of essential facilities and the risk of fire, as well as the collapse of outer walls, pipings or hanging balconies. These defects can lead to injury and loss of life. For example, an old woman was killed and several others injured when a pre-war building in Central District collapsed in 1991, while injuries resulting from the collapse of hanging balconies in Yaumati and Tai Kok Tsui District occurred in the same year.

The irregular or mixed land use of commercial, industrial and residential functions in one building or in one neighbourhood causes public nuisances of noise, fire risks, safety, security and heat in the building or neighbourhood. The lifts may be occupied for loading and unloading leaving no room for residents. Residents living by the side of cement plants, printing factories and textile factories are continuously affected by flying particles, noise and acid smell emissions.

The health of residents in urban slums is seriously affected by environmental pollution problems. For example, in Kennedy Town, the cattle depot, refuse incinerator, abattoir, cement plants, vegetable and poultry wholesale markets, lard and sharks' fin processing factories and open sewers create noise, smell, dust and other irritants. In October 1986, the Central and Western Development Research Centre carried out a lung function test and a questionnaire health survey for the residents of Kennedy Town. Not surprisingly the respondents reported a high incidence of chronic respiratory tract problems and asthma. This was con-

Table 8.2 Summary of environmental problems and low-income communities in Hong Kong

Low-income communities	Environmental problems and risks
Squatter areas	Fire risk, dangerous slopes, landslides, poor hygiene, lack of public toilets and standpipes, slippery paths, lack of public facilities, excessive heat.
Temporary housing areas	Remote, dangerous structures and decay, fire risk, excess heat, lack of public facilities and access unsafe.
Urban slums	Illegal constructions, falling balconies and outer walls, pollution from industrial buildings and public facilities which create nuisance, congestion and poor ventilation.

sistent with observations from local family doctors that children living near the pollutants suffered from coughs and respiratory tract infections more frequently than others. One hundred and twenty-eight residents were asked to blow as hard as possible into a peak-flow rate meter which records lung functional ability. The results showed that male residents exhibited on average a 100-litre per minute lower blowing rate than the standard rate and women exhibited on average an 80-litre per minute lower than standard rate. The lung functional capacity of the Kennedy Town residents was on average 30 per cent lower than the standard in Hong Kong (Hung, 1993).

Institutional and community responses

In the 1990s, the Hong Kong government and the public are all concerned about the problems of the environment. Three levels of response can be identified: the institutional and community responses at central planning level, departmental level and grassroots mobilisation level.

Central planning

During the period of *laissez-faire* 'active non-intervention' by the government in the sphere of industrial growth and development, pollution became a significant problem. Consequently, the government commissioned an overseas consultancy firm in 1974 to study existing conditions and to propose legislative guidelines and structural changes. As a result, a new Environmental Protection Unit (EPU), attached to the Environment Branch of the Government Secretariat was established in 1977. This developed into the Environmental Protection Agency (EPA) in 1981 and in 1986 became the Environmental Protection Department (EPD). The responsibilities of the EPD included policy-making, implementation and monitoring. In order to develop a comprehensive policy on environmental pollution, a policy white paper was published in 1989 which suggested that the government would deploy HK$20 billion in the coming decade to improve the environment. As an indication of the government's determination to protect the environment, a team of 950 staff have been employed to work on pollution control. In July 1989, the EPD was integrated into a new Planning, Environmental and Lands Branch of the Government Secretariat (see Figure 8.1). This move reflected the government's desire to link its urban planning more closely with environmental protection functions (Hills and Barron, 1990). For the major forms of pollution, planning is quite comprehensive, covering, for example, pollution processes, siting restrictions, emission controls and the treatment and disposal of waste. In

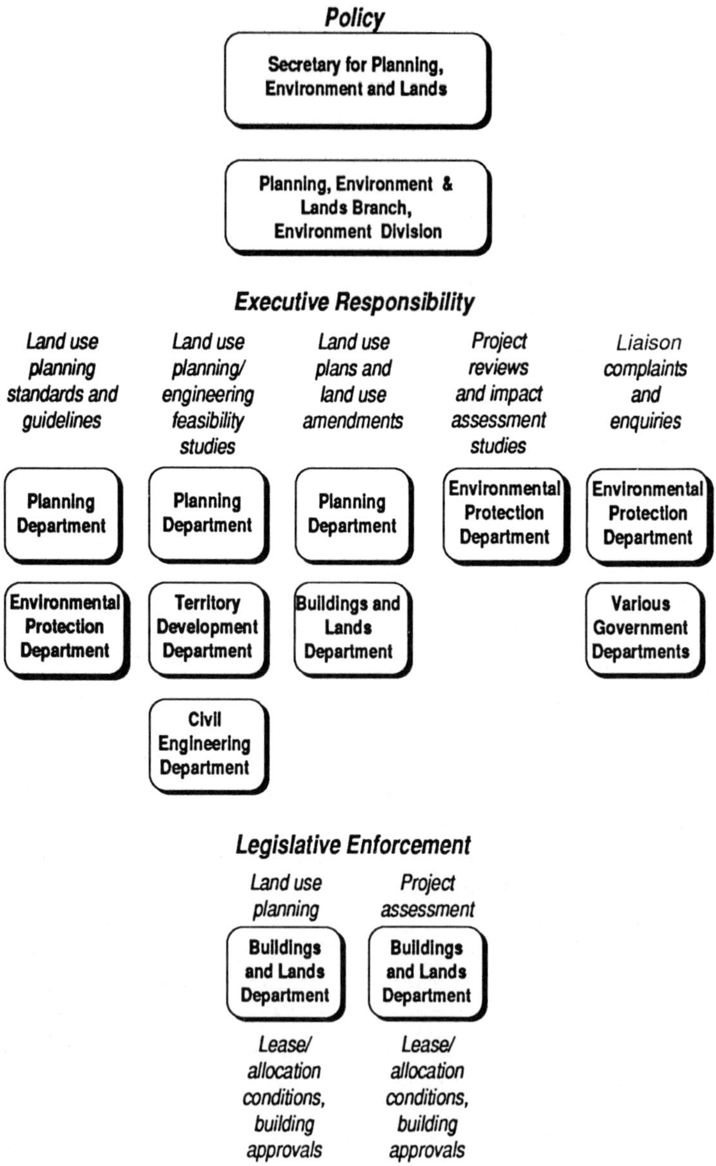

Figure 8.1 Government structures for environment, planning and assessment in Hong Kong. Source: Environment Protection Department, 1992, p. 28. Reproduced by permission

addition the Hong Kong Productivity Council has been assigned to provide assistance for factories to introduce affordable environmental protection measures. Such supportive advice is clearly essential with the enactment of controls on industrial gaseous emissions and waste-water discharge which mean that factories contravening these can now be heavily fined.

Central policy and law-making is the most effective mechanism for controlling pollution, and a list of relevant legislation adopted in Hong Kong can be found in Appendix 1. However, the legislative process is very slow following resistance from industrialists. Furthermore, controls focus on the prevention of further pollution but have relatively little effect on areas which have already been severely polluted. In reality this means that residents have to live with the pollution especially in old industrial areas such as Kwun Tong, Tsuen Wan and Kwai Chung. Participation of ordinary citizens in such policy matters is disappointingly low.

Departmental planning

Pollution control is disseminated through various government departments and public bodies. The most influential are the two Municipal Councils (Urban Council and Regional Council) and the Housing Department. The Municipal Councils and Urban/Regional Services Department control hawkers, run markets, inspect the hygiene of public eating facilities, and are also responsible for refuse collection and street cleaning. The general hygiene of the municipality depends upon the efficient operation of the Municipal Councils and Urban Services Department. Restaurants and cooked-food stalls are a major source of pollution, producing smoke and oily gaseous emissions, as well as noise from cooking and air-conditioning systems.

The Housing Department is playing a crucial role in determining the quality of the living environment for the poor in Hong Kong. The inflow of refugees from China in the late forties exacerbated the territory's housing problem. However, the government's public-housing programme did not start until 1954 following a disastrous squatter fire at Shek Kip Mei on Christmas night 1953 which rendered 53,000 people homeless. By 1990, 43.7 per cent of the population (513,849 households with 2,342,900 persons) and 39.5 per cent of the households were living in public-housing estates and 'Home-Ownership Scheme' housing (Census and Statistics Department, 1990, p. 187).

Public housing in Hong Kong is intended to accommodate households unable to afford self-contained flats in the private sector. While waiting for public housing, which always takes a few years, the applicants have to resort to other housing sources, mainly squatter areas and temporary housing areas. The median monthly household income levels for residents in temporary housing and squatter areas are among the lowest in Hong Kong, and as noted above, the living conditions in the squatter and temporary housing areas are far from satisfactory. The Housing Authority has made a number of attempts to improve its temporary housing areas in respect of their structure, layouts and amenity provisions, as well as to improve the environmental condition of squatter areas by initiating Squatter Area Improvement Programmes. The development of the squatter policy in Hong Kong will be discussed below to demonstrate how the government can contribute to systematic environmental improvement for the under-privileged group.

Squatter policy on environmental improvement

The squatter problem has existed in the territory for many decades. In the 1940s there were increasing incentives to squat on Crown Land as a result of the accelerated redevelopment of old buildings, the closure of dangerous properties and population growth resulting from immigrant influxes from China and natural increase (Hong Kong Government, 1964). Despite its long history, the squatter problem only became a pressing issue for the government in the early fifties when the squatter population reached a quarter of a million and large stretches of valuable land within, and at the fringe of, the urban areas were sterilised. However, the introduction of a Resettlement Programme in 1954 for squatters displaced by development and natural disasters was unable to arrest the growth of the squatter population. On the contrary, the number of squatters soared to about 600,000 in 1963 (Hong Kong Government, 1964) compared to a total population of 3.13 million (Pryor, 1983). The average annual rate of increase between 1954 and 1963 was about 35,000 (Choi, 1993).

Subsequent to the fierce blaze at Shek Kip Mei in 1953, a Resettlement Department was established to deal with all matters concerning the prevention of illegal structures, as well as the clearance and resettlement of squatters. It also marked the beginning of the government's involvement in public-housing provision. Between 1954 and 1975, a total of thirty-six multi-storey resettlement estates (Marks I–IV), comprising 471 blocks were completed (Hong Kong Housing Authority, 1975). The scale of the resettlement programme, however, could not keep pace with the growth in housing demand, which was boosted by waves of illegal immigrants from China. It was estimated that approximately 400,000 illegal immigrants had entered Hong Kong between 1978 and the withdrawal of the touch-base policy in October 1980 (Pryor, 1983).

Before 1982, the government's squatter policy was basically negative and reactive; resources were channelled towards squatter control and clearance. Despite the fact that the environmental conditions of squatter settlements were highly unsatisfactory, the government had no incentive to rectify the situation for fear of conferring on squatters a legal status which would encourage further squatting. A change in the government's squatter policy in the early eighties was attributed to the cancellation of a touch-base policy in late 1980 which had successfully stopped the massive illegal immigration inflow from China. A series of significant squatter fires and landslides in 1982 not only alarmed the government but also raised the public awareness of squatter problems. With the pressure from the Council on Squatter Policy, the government adopted a more positive approach to squatter problems and established a Squatter Areas Improvement (SAI) Division within the Housing Department in 1982 (Ng, 1993).

Prior to the introduction of the Squatter Areas Improvement programme, the living environment of squatter settlements had fallen well below acceptable standards. Basic amenities/facilities were either absent or inadequately provided. Public standpipes, for example, always fell short of demand, while a sewerage system was absent. Legal supply of electricity was not available until 1976, however illegal tapping of water and electricity through racketeers was widespread. Furthermore, access within squatter areas was poor and illumination at night was insufficient, while the provision of recreational and community facilities was unheard of. In addition, squatter areas were highly susceptible to fire and landslides.

The establishment of the Squatter Areas Improvement Division was intended to rectify these deplorable living conditions. The improvements included: provision of fire service inlets, fire mains, fire hydrants and installation of ten-metre fire breaks to reduce the fire risks; minor geotechnic work to reduce risk of landslide; as well as provision of water standpipes, street lights, toilets, refuse collection points, service channels and footpaths to improve the quality of life for the squatter residents. The SAI Scheme was a typical upgrading project, and the government covered most of the cost; the scheme ended in 1990 since most squatter areas were scheduled for demolition (Ng, 1993).

The District Boards are also allocated money for small-scale Local Environment Improvement Projects in the district. Before the implementation of the SAI Scheme, these small-scale projects were the only source for environmental improvement in squatter areas. With the termination of the SAI Scheme, the District Boards have had to resume their duty in upgrading of the environment in squatter areas. Since the resources and expertise of the District Board are limited, their attempt to improve the living environment in squatter areas is by no means a straightforward task. Clearly, public awareness and concern are crucial components in the success of these projects, thus the Central and Western District Board launched a 'Ten Good and Ten Bad Environmental Spots Nomination Competition' to arouse public concern and embarrass the relevant government departments (Hung, 1993).

Grassroots mobilisation

Given that urban policy broadly favours the rich, the poor have themselves taken on the task of improving their welfare by improving better services and enhancing economic opportunities (Yeung, 1990, p. 310; see also Yeung and McGee, 1986). Grassroots participation in environmental improvement is seen as a viable alternative to government intervention, perceived as organisationally efficient and cost-effective. Planners and governments have encouraged grassroots participation in low-income communities to improve their quality of life through collective action. In this way, the government commitment to service provision can be reduced and the quality of service can be improved. There are several government-sponsored grassroots mobilization schemes in Hong Kong such as the Neighbourhood-Level Community Development Project, the Mutual Aid Committee and Community Care Programmes.

Neighbourhood-Level Community Development Projects In 1977 the Hong Kong government endorsed the policy of setting up Neighbourhood-Level Community Development Projects (NLCDP) in 'deprived' communities which are defined by low income, lack of access to existing social services and lack of basic infrastructures. A team of three social workers is employed to work in deprived communities. To ensure that the government's investments are worthwhile these communities should not be demolished during the first three years. In 1991, there were fifty-one NLCDPs throughout the territories of Hong Kong (employing more than 150 social workers) engaged in the organisation of residents in squatter areas, temporary housing areas and urban slums. Residents are organised into task groups and residents' organisations to launch mutual help activities to improve the quality of life of the community.

Mutual Aid Committee and Owners' Corporation Multi-storey buildings are the commonest form of housing for the Hong Kong population. In order to mobilise residents or owners of domestic flats in multi-storey buildings to take part in the day-to-day administration of the buildings and prevent urban decay, the Hong Kong government enacted an Owners' Corporation (OC) Ordinance in 1971 and introduced a mutual help organisation called the Mutual Aid Committee (MAC) in 1973. The City and New Territories Administration offers MACs and OCs technical advice on building maintenance and provides the support to the community leaders which makes participation among the low-income population possible. MACs are given a quarterly stipend so that the organisation can pay its telephone and electricity bills. Unfortunately, there is a lack of leadership among residents in low-income neighbourhoods as a result of poor education and the need to work long hours. In this context, government involvement is essential in sustaining interest and momentum in environmental improvement projects and grassroots participation.

Community action and community care Besides NLCDPs, MACs and OCs, the government also sponsors community-centre services which organise citizens for mutual help. There is one community centre for every 100,000 people. The community centres organise women, children, youth, elderly people and residents' groups as well as launching community education programmes and community care activities to improve the quality of community living. In addition, there are other action-orientated voluntary agencies such as the Society of Community Organization (SoCO), Council on Public Housing Policy, the Council on Squatter Area Policy, and the Christian Industrial Committee (CIC) which are empowering the grassroots to fight for environmental improvement.

Thus SoCO organises tenants of 'cage' accommodations to fight for their housing rights, following demolition of their 'cage' apartment, and organises boat squatters in Yaumati to fight for their right to public housing. The Council on Public Housing Policy organises residents in public-housing estates (about half of the Hong Kong population) to bargain with the Housing Department for environment improvement. The Council on Squatter Policy mobilises squatters to collaborate in the implementation of SAI schemes, while the CIC mobilises workers to campaign for a safe working environment and organises victims of industrial accidents to seek compensation.

Clearly, environmental hazards are among the key issues which can attract public concern and involvement. Thus, there were mass mobilisations in Hong Kong in the 1980s against the Daya Bay Nuclear Plant, the refuse incinerator in Kennedy Town as well as the chemical waste and gas depot on the Tsing Yi Island (Hung, 1993; Chan, 1988; Ng, 1993). Since the mid-1980s, local politicians have been active in such grassroots mobilisation on environmental issues. Frequently, however, this involvement has been motivated by the need to build up political support rather than a concern for environmental issues *per se* (Hung, 1993a).

Issues of environmental protection in Hong Kong

The above review of central government policies, departmental programmes, local involvement and grassroots mobilisation projects highlights environmental interventions at various levels. Despite the apparently comprehensive multi-level approach towards environmental protection, there are problems with the Hong Kong system: policy bias towards the rich, the lack of cooperation between municipal departments, limited resident mobilisation and efficiency gaps.

Policy bias towards the rich

In many respects, one can identify a policy bias towards the rich in Hong Kong. Large corporations can and do exert a powerful influence over government legislature. Frequently, government attention is directed towards investment and industrial production instead of environmental protection; and the priority allocated to the latter is relatively low, despite statements to the contrary.

Large corporations in Hong Kong such as the Hong Kong Bank and the Swire Group have recently formed a 'Private-Sector Committee on the Environment' to contribute to environmental protection. While the committee removed rubbish from the Hong Kong Victoria Harbour to improve the public image of Hong Kong, this hardly tackles the key source of the pollution which is industrial-based (Menagh, 1991).

There are squatter areas which are heavily polluted by adjacent factories. In the case of the Diamond Hill Squatter Area, hot water from dyeing factories washes over the settlement every day. Very often there is no viable solution but to relocate the factory or the squatter population (Cheung, 1993). In most cases, the factories will not be relocated and the residents have to wait until the land is redeveloped before the government is willing to rehouse them, as rehousing depends on public land-use planning and not on hardship.

Inter-disciplinary cooperation in vain

Although there is some degree of division of labour in environmental protection at the various levels there is little inter-disciplinary cooperation. Most environmental protection policies are coordinated by the Environmental Protection Department which is dominated by professional planners and engineers. In reality, cooperation is manifest mainly in the form of the EPD explaining policies to the public and seeking cooperation from citizens during implementation of the projects. Communication and coordination between the relevant professionals is also limited. The existing planning structures do not encourage multi-disciplinary cooperation in the field of environmental protection, each professional group is relatively self-contained and only lip service is paid to the goals of inter-disciplinary cooperation and grassroots participation.

Residents and community organisers have never been regarded as part of the multi-disciplinary team in environmental protection. To a large extent they rely on their own motivation and commitment to acquire competence in community education and to mobilise the grassroots in support of environmental protection (Chan and Lam, 1991).

Limited residents' mobilisation and contribution

Owing to the technical nature of the problem, grassroots residents can easily be overwhelmed by chemical terms, consultancy reports and other highly sophisticated data. There are severe knowledge gaps among the public and there is insufficient resource support for the grassroots to bridge these and participate in environment-related issues. Therefore, the community organisers working in the NLCDPs have an important contribution to make in terms of consciousness-raising and training, thereby improving the competence of residents to be involved in the debate on problems and policies.

Effectiveness versus efficiency

Participation of the grassroots is frequently viewed as a time-consuming process hindering the efficiency of the environmental planning and management process. However, if the public are properly consulted during the process, their opinions can, and should, be adequately reflected in project planning. In short, citizen involvement can enhance the relevance and effectiveness of planning outcomes. In order that grassroots consultation will not unduly delay the planning process, it would be desirable if grassroots participation and consultation were institutionalised within the overall organisational framework.

Conclusion

Residents in low-income communities in particular are suffering from safety risks and environmental hazards over which they have little control. The government and the community have adopted a wide range of mechanisms to improve the housing conditions and the environment for the poor. The government has effectively housed half of the population in public housing and instituted the Squatter Area Improvement Scheme to offer site-and-services and NLCDPs to deprived squatter neighbourhoods. Politicians, voluntary organisations and residents' groups are also active in environmentally related issues. However, despite the effort of grassroots mobilisation, the major hazards of industrial pollution and chemical-waste disposal cannot be solved simply by mutual help among the low-income population. Nevertheless, community organisers can play an important role in raising the profile of grassroots groups and encouraging their potentially crucial involvement in the field of environmental protection. What is required, however, is a clearer documentation of the importance of consultation between the professionals and grassroots mobilisation (Chan and Hills, 1993) which would provide an important basis for future practices based on cooperation and a unity of purpose.

References

Asiaweek, 1991, 'Economics Indicators', 26 July 1991.

Bristow, R., 1984, *Land Use Planning in Hong Kong: History Policies and Procedures*, Hong Kong, Oxford University Press.

Census and Statistics Department, 1987, *Hong Kong 1986 Bi-Census*, Hong Kong, Hong Kong Government Printer.

—, 1990, *Annual Digest of Statistics*, Hong Kong, Census and Statistics Department.

Chan, C. and Hills, P. (eds), 1993, *Limited Gains: Grassroots Mobilisation and the Environment in Hong Kong*, Hong Kong, Centre of Urban Planning and Environmental Management, University of Hong Kong.

Chan, C. and Lam, D., 1991, 'A review of models of attitude and behaviour change in children and youth service: promotion of environmental concerns', in *Children and Youth Service Models in Perspective 1991*, Hong Kong, Hong Kong Council of Social Service.

Chan, Cecilia, 1980, 'Community development through physical improvements', *Community Development Resource Book 1979–1980*, Hong Kong Council of Social Service, pp. 42–8.

—, 1981, 'An exploration into the squatter-area problems in Hong Kong', *Social Service Quarterly*, **79**, pp. 33–6 (in Chinese).

—, 1984, *Towards Theory Building for the Neighbourhood Community Development Practice in Hong Kong*, unpublished dissertation for M.Soc.Sc., University of Hong Kong.

—, 1988, 'Citizen participation in town planning: the case of Hong Kong', in *Casebook of Social Work Intervention 1988*, Hong Kong Council of Social Service, pp. 98–117.

Chan, W.K., 1991, 'Environment and social development: the case of Hong Kong', ICSW 1991 Asia and Pacific Regional Conference, *Environmental Protection and Social development: Role of Non-Government Organizations. Conference proceedings (Vol. 1)*, Hong Kong Council of Social Services, pp. 18–27.

Cheung, R., 1993, 'The campaign approach in grassroots mobilization', in C. Chan and P. Hills (eds), *Limited Gains: Grassroots Mobilisation and the Environment in Hong Kong*, Hong Kong, Centre of Urban Planning and Environmental Managment, University of Hong Kong, pp. 187–94.

Choi, P., 1993, 'Environmental protection in Hong Kong: an historical account', in C. Chan and P. Hills (eds), *Limited Gains: Grassroots Mobilisation and the Environment in Hong Kong*, Hong Kong, Centre of Urban Planning and Environmental Management, University of Hong Kong, pp. 83–98.

Douglass, M. 1992, 'The political economy of urban poverty and environmental management in Asia: access, empowerment and community-based alternatives', *Environment and Urbanisation*, **4** (2), pp. 9–320.

Environmental Protection Department, 1990, *Environment Hong Kong 1990: A Review of 1989*, Hong Kong, Hong Kong Government Printer.

—, 1992, *Environment Hong Kong 1992: A Review of 1991*, Hong Kong, Hong Kong Government Printer.

Environmental Protection Department and Planning Department, 1991, *Environmental Guidelines for Planning in Hong Kong*, Hong Kong, Hong Kong Government Printer.

Government Information Services, 1991, *Hong Kong 1991: a Review of 1990*, Hong Kong, Hong Kong Government Printer.

Governor's Speech, 1982, Hong Kong, Government Printer.

Hills, Peter and Barron, William, 1990, 'Hong Kong: can the dragon clean its nest?', *Environment*, **32**, (8), pp. 17–45.

HKYWCA and HKUGA, 1980, *Research Report of the Community Health Study in the Hoi Pong Village*, Hong Kong, the Hong Kong Young Women's Christian Association and the Hong Kong University Graduates' Association.

Hong Kong Government, 1964, *Review of Policies for Squatter Control, Resettlement and Government Low-Cost Housing*, Hong Kong; Hong Kong Government Printer.

Hong Kong Housing Authority, 1975, *Hong Kong Housing Authority Annual Report 1974/75*, Hong Kong, Hong Kong Government Printer.

Hung, W.T., 1993, 'The role of District Board in community-based environmental management', in C. Chan and P. Hills (eds), *Limited Gains: Grassroots Mobilisation and the Environment in Hong Kong*, Hong Kong, Centre of Urban Planning and Environmental Management, University of Hong Kong, pp. 51–62.

—, 1993a, 'Politicalization of the environment and the role of non-governmental organizations', in C. Chan and P. Hills (eds), *Limited Gains: Grassroots Mobilisation and the Environment in Hong Kong*, Hong Kong, Centre of Urban Planning and Environmental Management, University of Hong Kong, pp. 41–50.

Lai, M.Y., 1983, *In-Situ Environmental Improvement in Urban Squatters – An Expedient Solution?*, unpublished M.Soc. Sc. Dissertation, University of Hong Kong.

Lee, James, 1991, 'Housing studies, urban theory and the living environment: the case of Hong Kong'. ICSW 1991 Asia and Pacific Regional Conference, *Environmental Protection and Social Development: Role of Non-Governmental Organizations, Conference Proceedings (Vol. 1)*, pp. 156–68.

Menagh, M., 1991, 'The greening of management companies worldwide have begun to protect the environment in a multitude of ways', *Newsweek*, 17 June.

Newsweek, 1991, 'Special Advertising Section – The Environment: A Corporate Challenge', 17 June.

Ng, Cho-nam and Ng, Ting-Leung, 1992, 'The environment', in J.Y.S Cheng and P.C.K. Kwong (eds) *The Otner Hong Kong Report 1992*, Hong Kong, Chinese University Press.

Ng, H.S., 1993, 'The Tsing Yi residents' mobilization against environmental hazards, in C. Chan and P. Hills (eds), *Limited Gains: Grassroots Mobilisation and the Environment in Hong Kong*, Hong Kong, Centre of Urban Planning and Environmental Management, University of Hong Kong, pp. 63–82.

Ng, Yiu-fai, 1993, 'Environment and grassroots participation in squatter areas', in C. Chan and P. Hills (eds), *Limited Gains: Grassroots Mobilisation and the Environment in Hong Kong*, Hong Kong, Centre of Urban Planning and Environment Management, University of Hong Kong, pp. 99–115.

Pryor, E.G., 1983, *Housing in Hong Kong*, Hong Kong, Oxford University Press.

Wong, S.K., 1991, 'Efficiency versus rights: resource allocation and the environmental protection policies', *Journal of Policy Viewers*, May, pp. 23–9.

Yeung, Yue-man, 1990, *Changing Cities of Pacific Asia: A Scholarly Interpretation*, Hong Kong, The Chinese University Press.

Yeung, Y.M. and McGee, T.G. (eds), 1986, *Community Participation in Delivering Urban Services in Asia*, Ottawa, International Research Centre.

Appendix 1

Hong Kong Legislation and Related Events Concerning Environmental Control (1974–90)

1974 Formation of the Advisory Committee on Environmental Pollution (EPCOM); the whole territory became a smoke-control area; Environmental Resources Ltd commissioned to undertake a comprehensive investigation into the growing need for environmental control in Hong Kong.

1975 Merchant shipping (Oil Pollution) Ordinance makes oil pollution liability insurance mandatory for all cargo-carrying vessels.

1976 Public Health and Urban Services (Minor amendments) Ordinance prescribes acceptable levels of air-conditioner noise. Factories and Industrial Undertakings Ordinance regulates permissible noise levels; Establishment of Environmental Protection Unit (EPU), within the Environmental Branch (based on ERL's preliminary report), as a small unit responsible for developing policy and formulating legislation on environmental protection.

1977 Ban on aircraft movements between midnight and 6.30 a.m.; ERL completes study and submits proposed noise abatement, waste-disposal, water pollution control, air pollution control and environmental impact ordinance.

1978 Summary Offences (Night Work) Ordinance puts limits on night-time construction work noise; government replaces 1974 Advisory committee on Environmental Pollution with an Environmental Protection Advisory Committee.

1980 Water Pollution control Ordinance introduces water control zones; Enactment of the Waste Disposal ordinance provides statutory powers over waste collection and disposal for the relevant authorities.

1981 EPI detaches from the Government Secretariat to form the Environmental Protection Agency (EPA); Environmental Branch is dissolved, EPA supervised by the Home Affairs Branch.

1982 Tolo Harbour declared as the first Water Control Zone (WCZ); EPA is placed under the Health and Social Welfare Branch.

1983 Air Pollution Control Ordinance (APCO) introduces air-control zones and air quality objectives in respect of these zones, enforced by the Air Pollution Control Division (APCD) of the Labour Department.

1984 Government replaces 1978 Environmental Protection Advisory Committee with an Environmental Pollution Advisory Committee.

1985 Draft Water Pollution Control (WPC). Regulations under public consultation.

1986 EPA is restructured to EPD to centralise all pollution control units with government; WPC regulation was passed in Legislative Council.

1987 First Appointed Day is announced on Tolo Harbour.

1988 Southern WCZ is announced. Sewerage Master Plans and Livestock Waste Control Scheme are implemented; Noise Control Ordinance is enacted; EPD is transferred to the newly established Planning, Environment and Lands Branch.

1989 The White Paper on Pollution – *A Time to Act* is published; Port Shelter and Junk Bay WCZs are announced; a Waste Disposal Plan is published; Sewage Disposal Strategy was adopted; Ozone Layer Protection Ordinance (1989) is enacted.

1990 Deep Bay and Mirs Bay are declared WCZs; the first refuse transfer station at Kowloon Bay becomes operational; Lai Chi Kok incinerator and Jordon Valley Landfill are phased out; Environmental Campaign Committee is established and the first Environmental Festival is launched.

9 Marginal urban environments as havens for low-income housing: Third World regional comparisons

Hamish Main and Stephen Wyn Williams

Third World urbanisation has produced a number of unexpected outcomes in recent decades, notably the rapidity of urban growth overall and the mega-populations of the largest cities. This chapter derives from another, related, feature of Third World urbanisation that was unforeseen by social scientists: the spate of disasters that has befallen cities around the Third World, especially since the 1970s. Such urban disasters are clearly a part of the wide range of environment-related calamities suffered by Third World people, but the apparently enhanced vulnerability of Third World cities to natural environmental hazards arises principally from the residential occupation of marginal urban environments by those who are unable to secure residence in less vulnerable sites. It is argued below that this is an outcome of the operation of urban land and housing markets, as these markets develop in Third World societies in ways that increasingly favour those with sufficient wealth and marginalise those without. Unable to afford accommodation in preferred sites, the urban poor become increasingly constrained to accommodation on land not wanted by the better-off.

The most visible result of residential marginalisation has been the concentration of legal and illegal low-income housing towards the urban periphery, especially in Latin America. But when journeys to the city centre become too long, or when land at the periphery becomes scarce, the urban poor begin to look for inner-city sites previously discounted as unavailable not because they are required for the better-off but specifically because they are *not* required due to their awkward or dangerous nature.

Thus, when a Third World city's land and housing markets are sufficiently developed, the chief locational determinant of low-income housing is the attribute of low evaluation by middle- and high-income housing markets and by non-residential land markets. The urban poor obtain shelter on land, left over from other markets, that is marginal either in a spatial sense (locationally peripheral) or in an environmental sense (of intrinsically poor site quality).

But the development of land and housing markets is a lengthy and uneven process in Third World societies, involving squatter demolitions as the stick where land is required for others' legal purposes, as well as the carrot of cheap land and/or housing where competition is less severe and the poor can find safe residential havens. The market guides relegation of the urban poor into marginal residential havens – all too often dominated by negative externalities and peripherality, yet havens from the predations of formal land and housing markets supported by the state. This process is less advanced overall in Africa than in Latin America, with much of Asia being in an intermediate position.

This is not a novel explanation; these issues have been considered *inter alia* by Angel *et al.* (1983); Baross (1983); Gilbert and Ward (1985); Doebele (1987); Hardoy and Satterthwaite (1989); and Baross and van der Linden (1990). It is applied in this chapter to a comparison of Latin American and African cities. Such a broad regional comparison is by no means intended to suggest that differences between cities and urban cultures within each continent are insignificant, but we believe that sufficient regional similarities are discernible to make such an exercise worthwhile. Wealthier and more powerful urban residents tend to reside in a city's more desirable sites in most societies, pre-industrial as well as industrial, but recent urban environmental disasters have been sufficiently frequent and damaging to sufficient numbers of urban residents to warrant close scrutiny of the forces acting upon those who are worse affected in Third World cities towards the end of the twentieth century.

The chapter begins with a brief survey of what constitutes marginal urban environments. It may appear at first glance that Latin American cities have to cope with more difficult topographic and other *in situ* conditions than do African cities, but closer examination reveals site problems of one kind or another in almost every urban area in Africa. Yet it is apparent that the urban poor in Latin American cities today are constrained to reside in these marginal environments to an extent markedly greater than in the case of the African urban poor. It is therefore unsurprising that the list of environmental disasters which have recently befallen the poor in so many Latin American cities is longer than in the African case. The chapter proceeds by offering a model of urban residential development which indicates spatially how the urban poor are constrained to marginal environments and by briefly considering some aspects of the development of land and housing markets and associated processes which bring about this situation. The logic of the model is that as African societies become increasingly urbanised, as African cities grow in population and in area and as their land and housing markets develop, they will in coming decades be increasingly threatened by environmental disasters that wreak widespread damage to homes and to lives.

Marginal urban environments

Marginal urban environments are sited in and/or around negative externalities. These negative externalities are natural or man-made features of urban environments that make nearby residence unattractive because they entail actual or potential ongoing problems for local residents and/or threats of disaster (see Table 9.1). The most widespread features of the natural environment are probably steep slopes, which make access and construction difficult and may threaten landslides; flood plains, swamps and other water margins, which may cause problems for access and construction and health, and may threaten flood; and zones threatened by tectonic activity. Industrial production sites are probably the most problematic man-made feature of urban environments, with their threat of environmental pollution or even major disaster, but rubbish dumps, burial grounds, borrow pits, airport flight-paths and high-tension electric cables also figure prominently in creating undesirable residential locations in Third World cities.

Where a dominant wind direction is more or less constant throughout the year, residential areas downwind of noxious urban activities are readily identifiable as marginal environments; but it is more usual for variation in wind direction to reduce the role of wind pollution in determining which urban locations are environmentally marginal for desirable residence. Downstream sites, drawing water that has already passed through much of the urban system, are more commonly identifiable as marginal, given the health risks posed by polluted water in Third World cities: a cholera outbreak in Jakarta in August 1977, for example, was traced to spring water polluted by waste infiltrated upstream. Zones of tectonic activity are perhaps the least site-specific of negative externalities in their effects on residential areas. But, as Table 9.1. suggests, the areas of the city that are affected by most negative externalities can be identified fairly closely.

Few cities are located on anything like the featureless isotropic plain of the theoretical models. Buenos Aires is unusual among major Latin American cities in its flat site. Africa appears at first glance to offer more sites for unconstrained urban expansion than other Third World continents: the monotonous plateau of Zambia's Copperbelt towns is interrupted less by natural features than by open-cast mines and their waste dumps, while Casablanca's 'topography is permissibly favorable, rising as it does gently inland from the ocean front' (Hance, 1970, p. 306). But actual urban development may provoke an apparently generous topography into

Table 9.1 Negative externalities and threats of disaster as determinants of marginal urban residential environments

	Ongoing problem		Threat of disaster	
Negative externality	*Problem*	*Site specificity within city*	*Disaster*	*Site specificity within city*
Tectonic zone	construction	low	earthquake/ volcano	low
Steep slope	construction access	very high very high	landslide	very high
Swampy land	construction access insects/disease	very high very high high	flooding epidemic	high high
Floodable land	insects/disease construction access	very high very high very high	epidemic flooding	very high very high
Industrial production zone	airborne pollution waterborne pollution noise crowds of workers	high/low high very high high	gas leak/ explosion epidemic	very high high
Rubbish dump	airborne pollution waterborne pollution insects/disease	high/low high high	methane explosion epidemic	very high high
Burial ground	waterborne pollution superstitious fears	high high	epidemic	high
Borrow pit	insects/disease construction	high very high	epidemic	high
Airport flight path	noise	high	crash	high
High-tension electricity cables	noise/illness	very high	electrocution	very high
Railway line, main road	airborne pollution noise	very high very high	crash	very high
Sewer	waterborne pollution insects/disease	high high	gas leak/ explosion epidemic	high high

difficult or disastrous consequences, if development plans are not sufficiently thorough. Suez's superficially easy terrain encouraged imaginative geometrical urban development plans for high-density housing – until geomophologists identified a substructure of very steep slopes, underground wadis constituting flood hazard, a high water table and very saline soils that would attack concrete foundations (Doornkamp, 1985, pp. 76–7). Development constraint maps in the town plans of reliable survey and planning companies may show severely limited directions for the desirable future expansion of even small towns on extensive plains, though all too often less thorough planning surveys do not identify flood threats, poorly drained soils, steep slopes and other potential physical constraints.

Clearly, one person's negative externality may be another's positive externality. It was common in pre-industrial Islamic cities, for example, for noxious activities like dyeing and tanning to be confined to locations outside the city wall, or at least towards the periphery of the built-up area; these were negative externalities for the majority of the urban population, but not so for the dyers and tanners themselves who resided close to their workplaces. Similarly in the industrial city, factory employees without reliable transport have a reason to seek accommodation close to their workplaces despite the pollution suffered there. Lagos is unusual in that some of the low-density residential areas surrounding the city centre are closer to major industrial areas (for example at Apapa) than are poorer neighbourhoods. This is partly a function of the awkward island site, reflecting especially the difficulties of transport between the centre and periphery of Lagos. Transport routes may be considered either positive or negative externalities; but low-income neighbourhoods under airport flight paths (for example Kano, Lima) and alongside railway tracks (for example Delhi, Kuala Lumpur) and precarious residence on street pavements (for example Bombay, Calcutta) are clearly undesirable sites in major respects.

The quality of urban environments is perceived to some extent through the filter of culture, and such perceptions may change through time. In Britain council housing became a major negative externality for private housing markets between the world wars, for example, as did many major football grounds during the 1970s as a result of football hooliganism. The question of whether hillsides are desirable residential locations depends only partly on their topography and geology (accessibility and construction quality), and it is complicated by altitude. Europeans colonising tropical lands often expressed their preference for cooler climates by locating their centres of administration at higher altitudes. This was most evident in Spanish America (for example Bogota, Caracas, San Jose), where in places it followed the ancient precedent of the Maya and Inca cities. In India the colonial summer capital was located for some years in the Himalayas at Shimla, but such sites were more often used by colonials for the establishment of hill stations for recreation (for example Darjeeling, Baguio City, Yogjakarta) (King, 1991). Today, hillside residential sites are dominated by the wealthier members of many North American cities, while flat lowlands are left over for the poor. But this situation is reversed in Latin American cities like Rio de Janeiro and Caracas, Bogota and Lima, where it is predominantly the poor who reside on the hillsides. The altitude of La Paz is sufficient for those born in the city to prefer the lower hill slopes to the flat land of the higher El Alto above 4,000 metres, whereas migrants from the Altiplano, accustomed to harsh climatic conditions there, are now moving to El Alto in large numbers (van Lindert, 1991, p. 455). Those who resided until the mid-twentieth century in inner-city mansions in many Latin American societies moved out in large numbers to new peripheral suburbs as industrial development began close to city centres and when the increasing availability of private motor transport made suburbs more accessible. This automobile-encouraged surbanisation took a unique form in Salvador, where the elite had for centuries resided at the summit of the escarpment, leaving the flood-prone valley floors for gardening while the poor came to occupy the upper and middle hill slopes. In the 1950s, with the construction of motor roads along the coast, the elite descended *en masse* to new high-rise flats along the ocean shore, while the growing volume of *favelas* occupied not only the steeper hill slopes but also flood-prone locations alongside coastal streams and lagoons (Claval, 1991, p. 24) and even islands of garbage in the bay (Dwyer, 1975, p. 32).

The results of a brief survey of recent published literature, on the presence or absence of some major negative externalities and of major environmental disasters, are presented in Table 9.2. Twenty large cities in each of Latin America, Africa and southern Asia are listed in the table, which omits cities for which no such sites or events have been encountered in the literature; but such cities are few, especially in Latin America. The brevity of the survey and the interim nature of these results are stressed, and conclusions drawn from the table must be interpreted in this light. The table suggests that the volume of negative externalities is not dissimilar in each of the three regions surveyed, but that there are clear differences between the three in the predominant type of major negative externality: steep slopes are by far the most apparent in Latin American cities, whereas industrial production zones are most ubiquitous in southern Asian cities, and in African cities both of these appear to be about as widespread as swampy land. But it appears that urban environmental disasters arising out of negative externalities have been far more numerous in Latin America than in Africa, which has experienced far fewer earthquakes and landslides and industrial disasters and slightly fewer urban floods. The Asian cities surveyed appear to have been intermediate in this respect.

We have also sought to identify where low-income neighbourhoods have been markedly more affected by negative externalities than have middle- and high-income neighbourhoods, and preliminary findings appear in Table 9.2. Low-income neighbourhoods do indeed appear to have been much worse affected in a great majority of the cities surveyed in Latin America, but this is much less apparent in the African cases encountered. Steep slopes and swampy land are evidently inhabited by better-off residents in many African cities as well as by those less well off, though industrial production zones which are reported as major negative externalities are clearly associated far more with low-income housing. And, unlike Latin America, rarely has it been reported in African cities that environmental disasters have affected the urban poor markedly more than better-off urbanites.

Each continent has its notorious industrial areas: Sao Paulo's Cubatao ('Pollution Valley'), Rio's Belford Roxo and some of the eastern districts of Mexico City; Jakarta's Cibubur and the east coast of Bombay Island; the industrial districts of southern Tunis and south-western Alexandria. In each of these cases, the factories are surrounded by low-income neighbourhoods. Neither these residents, nor those who can afford to reside on the other side of the city, know what dangers exist in the factories or what disasters they might create. In some cases, air pollution is suffered by all urban residents more-or-less regardless of class or location; but neighbourhoods downstream from industrial zones, with their water pollution problems, are more likely to house low-income people. And when disaster strikes, it is the local neighbourhoods and their residents that almost inevitably suffer most. The escape of gas from Union Carbide's Bhopal plant in December 1984 killed perhaps 2,500 people mostly in neighbouring high-density *bastis*, while middle- and upper-class neighbourhoods on the other side of the city were largely unaffected (Jones, 1988, p. 58). The strong spatial correlation between areas affected by the gas and the city's high-density residential areas is clear; Bhopal's low-density residential areas are located very largely on the other side of the city centre, south of the Lower Lake. It was not only the *bastis'* location, but also the nature of their construction, that brought disaster out of hazard: those fortunate enough to stay on the upper floors of multi-storey buildings, or inside well-sealed buildings, suffered markedly less (Weir, 1986, p. 3). The explosion from a leaking petrol pipeline in Cubatao, Sao Paulo in February 1984 set off fires in a nearby shanty town and killed over 500 people (Weir, 1986, p. 115).

It is striking that only one of the thirteen industrial disasters indicated in Table 9.2 has occurred in any of the African cities listed. Flooding appears to have been the greatest urban environmental threat in Africa: it has become an almost annual hazard for some of the residents of Nigerian cities like Lagos and Port Harcourt (see Peil's Chapter Ten in this volume for example). But urban flood damage here has so far been minor compared with that faced by Khartoum residents in August 1988, when heavy rains combined with the annual Nile flood to wash away entire neighbourhoods, many of them temporary settlements of refugees from the war in southern Sudan (Ahmad, 1989, p. 54). A government source estimated that over 800,000 of Khartoum's population of about 2,500,000 had been 'affected' by the floods, though no number of lives lost was given

Table 9.2 Major negative externalities and environmental disasters in selected large Third World cities

City	*Major negative externality*			*Disaster*			
	Steep slope	Swampy or floodable land	Industrial pollution	Earth-quake	Land-slide	Flood	Industrial
Latin America							
Bogota	*		*		*	*	
Caracas	*		*	*	*		x
Cartagena		*	x				
Guatemala City	*		x	*	*	x	
Guayaquil		*				x	
La Paz	*				*		
Lima	*		*	x			x
Managua	x		x	x	x	x	
Medellin	*				*		
Mexico City	*	*	*	*		*	*
Quito	*	*					
Recife	*	*				x	
Rio de Janeiro	*	*	*		*	*	
Salvador	*	*				*	
San Jose	*			x	*	*	
San Juan		*				*	x
San Salvador	*			*	x	*	
Santiago	x		x				
Sao Paulo		*	*				*
Tijuana						*	
Africa							
Abidjan	x	*	*				
Accra	*	x	x	x		x	
Addis Ababa	x				x		
Alexandria			*				x
Cairo			*	*		x	
Dar es Salaam		*	*			x	
Douala		*	*				
Johannesberg		*	*		*	*	
Kampala		*				x	
Khartoum		x				*	
Kumasi	x	x					
Lagos		x	x			*	
Luanda	x				x		
Maputo	*	*					
Monrovia		*	*				
Nairobi	x	*	*				
N'Djamena		x			x	x	
Port Harcourt		*				*	
Rabat	*	*					
Yaounde		*	*				

Table 9.2 (*continued*)

	Major negative externality			*Disaster*			
City	Steep slope	Swampy or floodable land	Industrial pollution	Earth-quake	Land-slide	Flood	Industrial
Asia							
Ankara	*				*		
Baguio City	*	*		x	x	*	
Bangkok		*	*			*	x
Bhopal			*				*
Bombay		*	*			*	x
Calcutta		*	x			*	
Delhi		*	x			*	*
Hong Kong	*	*			*	x	
Jakarta	*	*	*			*	x
Karachi		*	*				
Kuala Lumpur	*	x	x			*	x
Madras		*	*				
Manila		*	x			*	
Pusan	*						
Seoul	*		x				
Shanghai			x				x
Surabaya		*				*	
Taichung			*				
Tehran			x				
Tel Aviv	x			x		x	

x Given externality is present/disaster has occurred, but no evidence found that low-income housing areas affected markedly more than other housing areas.
* Given externality is present/disaster has occurred, and low-income housing areas affected markedly more than other housing areas.
NB The twenty large cities in each of the Third World regions selected for Table 9.2 are the twenty for which the most information on major negative externalities and disasters has been encountered in the brief survey of recent literature undertaken for this chapter.

Sources: various, including Clarke *et al*, 1989; Douglas, 1983; Hance, 1970; Jones, 1988; Susman *et al.*, 1983; *Environment and Urbanization*, 1989a; *Environment and Urbanization*, 1989b; *Regional Development Dialogue*, 1991

(Ministry of Social Affairs, 1989; quoted by Abu Sin and Harbi, 1991, p. 123). Even though the 1988 Nile flood was lower than that of 1946, the devastation caused was estimated to be ten times greater due to housing encroaching on to unsuitable sites. Squatter settlements without drainage or open spaces or building regulations had grown very rapidly in Khartoum during the 1980s, accounting by the end of the decade for over 40 per cent of the built-up area; at the same time, 50 per cent of this area was 'exposed with its occupants to the risk of disaster from any abnormalities in Nile level and rainfall' (Abu Sin and Harbi, 1991, p. 121).

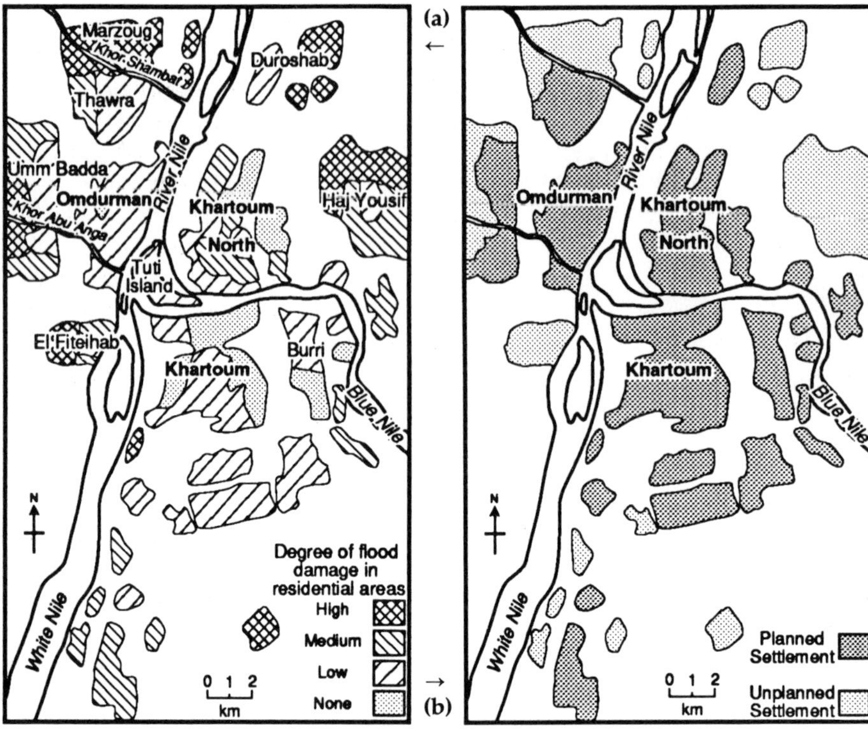

Figure 9.1 (a) Flood damage in greater Khartoum, August 1988. Source: after Abu Sin and Harbi, 1991; (b) Planned and unplanned settlements in Khartoum. Source: after Ahmad, 1989, p. 51

A comparison of Figure 9.1(a) with Figure 9.1(b) confirms that unplanned settlements around Khartoum's periphery took the brunt of the damage.

One of the squatter settlements most damaged by the 1988 Khartoum flood was Hilat Kusha ('Place of Rubbish'), a dump covering about one square mile sited next to the city's largest industrial estate, where over 5,000 refugees from the south dwelt among some of the factories' worst by-products (Save the Children Fund, 1990, pp. 36–7). Multiple negative externalities are not unusual in Third World cities, but this was a particularly abysmal combination. About 20,000 residents, many of them long established, were reported (Jimenez and Valasquez, 1989, p. 53) around 'Smokey Mountain', a Manila rubbish dump where many earn a living by scavenging garbage. Immigrants to Beirut used to be quarantined in Quarantina, subsequently abandoned and declared by the government as unsuitable for housing because it was adjacent to the municipal rubbish dump; squatters were sufficiently established by the 1970s to be renting accommodation here to Kurds and other newcomers (UNCHS, 1987, p. 140). Almost 40,000 people dwelt during the 1950s in Asima, an area of wasteland and rubbish dump in Baghdad used previously as a source of building mud and clay (Dwyer, 1975, p. 4). The borrow pits yielding material created by digging for material are potentially negative externalities wherever clay and mud are important in construction, as in much of central and southern Asia and the Andes (Agarwal, 1990, p. 35). Borrow pits for the construction and maintenance of city walls around Kano and Zaria and other northern Nigerian cities prior to the colonial era, as well as for housing today, have now been incorporated into the expanding built-up area; rainy-season inundation turns them into mosquito breeding grounds, and the health of people in adjacent neighbourhoods suffers accordingly. Residents of marginal settlements sited under high-tension electricity cables in Surabaya pay high rental costs despite their illegality and the overhead threat because they are located near centres of employment (Silas, 1989, pp. 65–6).

The 1988 Khartoum flood disaster illustrates the maxim that hazards are natural, whereas disasters are man-made. Well-organised urban development would avoid such disasters chiefly by diverting residential expansion to land not threatened by floodwaters, which after all are a recurring element of the Khartoum landscape. The question of why urban development is in fact not well organised here and elsewhere is complex; in keeping with much of the recent literature on the subject, we point in the second half of this chapter to market forces within capitalist development as the principal motor. A starting point for this explanation lies in the assumption that, as Table 9.2 indicates, urban environmental disasters, even of a magnitude much smaller than that of the 1988 Khartoum floods, have been far rarer in Africa than in Latin America.

Urban spatial development and marginal urban environments

Baross (1983) argues that the supply of land for low-income groups divides into three main categories: non-commercial, commercial and administrative. Within this broad classification marginal urban environments fall predominantly into the first of these, in so far as a monetary transfer price (at least initially) is not involved. According to Baross the settling of marginal land is a by-product of the increasing commercial articulation of land supply, and two basic reasons for its utilisation are posited. First, when no land is available from other non-commercial sources or administrative measures, 'those who cannot pay for land at all (not even for the cheaper rents) have no other option but to seek out parcels where nobody else wants to build' (p. 192). Second, where marginal land is centrally located it provides ready access to city-centre employment opportunities, thus 'if for topographic or land-use planning reasons marginal land exists near the city centre, these become strategic locations for popular development' (p. 192). However, this supply of 'free' land can be eroded over time as the established settlers start selling plots, becoming in effect petty landlords. In some countries, a greater threat to the continued use of marginal land is posed by the state, whose efficiency in administering master plans, and applying stricter zoning by-laws and building regulations prohibits the creation of new settlements and encourages the demolition of existing ones. However, in other countries the state's influence is much less pronounced.

With these basic ideas in mind it is possible to construct a relatively simple dynamic 'model' which illustrates over time how marginal urban environments acquire significance in the housing strategies of low-income groups. The emergence and expansion of marginal urban environments is presented in Figure 9.2, a three-stage model of urban development based on Latin America in the twentieth century. Three types of negative externality are evident in the model: (a) a river and its banks, which are unsuitable for housing construction and liable to flood; (b) steep slopes, difficult to access and to build on, and potentially provoked into landslides by such human intervention; and (c) an industrial zone, developing after the first stage of the model, which emits airborne pollutants especially in the direction of the prevailing winds towards the south and east. The model focuses on the spatial pattern of low-income residential areas *vis-à-vis* other residential areas and these negative externalities.

The first stage of the model is characteristic of Latin American cities prior to 1930, when motor transport was unimportant and most place-to-place movement was on foot. Wealthier urban residents lived close to the city centre, while the urban poor (many of whom worked as domestic servants in the mansions) resided mainly in rented accommodation or on communal land further from the centre. City population was small, and intra-urban distances sufficiently short for housing needs to be met from available land without major problems. Hardly anybody resided on the riverbank or on the steep slopes.

Industrial development, and the expansion of private transport for the better-off along with public transport for the urban masses, are incorporated into the second stage of the model. This typically represents a date after the middle of the twentieth century, perhaps about the 1960s. Wealthier residents have moved out of the urban centre some decades previously into extensive, well-serviced suburbs. Their former mansions have been subdivided and inhabited by lower-income residents, but most low-income neighbourhoods are located beyond the comfortable suburbs in poorly serviced shanty towns. Pressures on land are reflected in the variety of legal and illegal forms of access to land, in the incorporation of former communal land into the commercial sector and in evictions of squatters to make way for legal suburban development. Difficulties of access to the city centre and other major sources of employment, notably

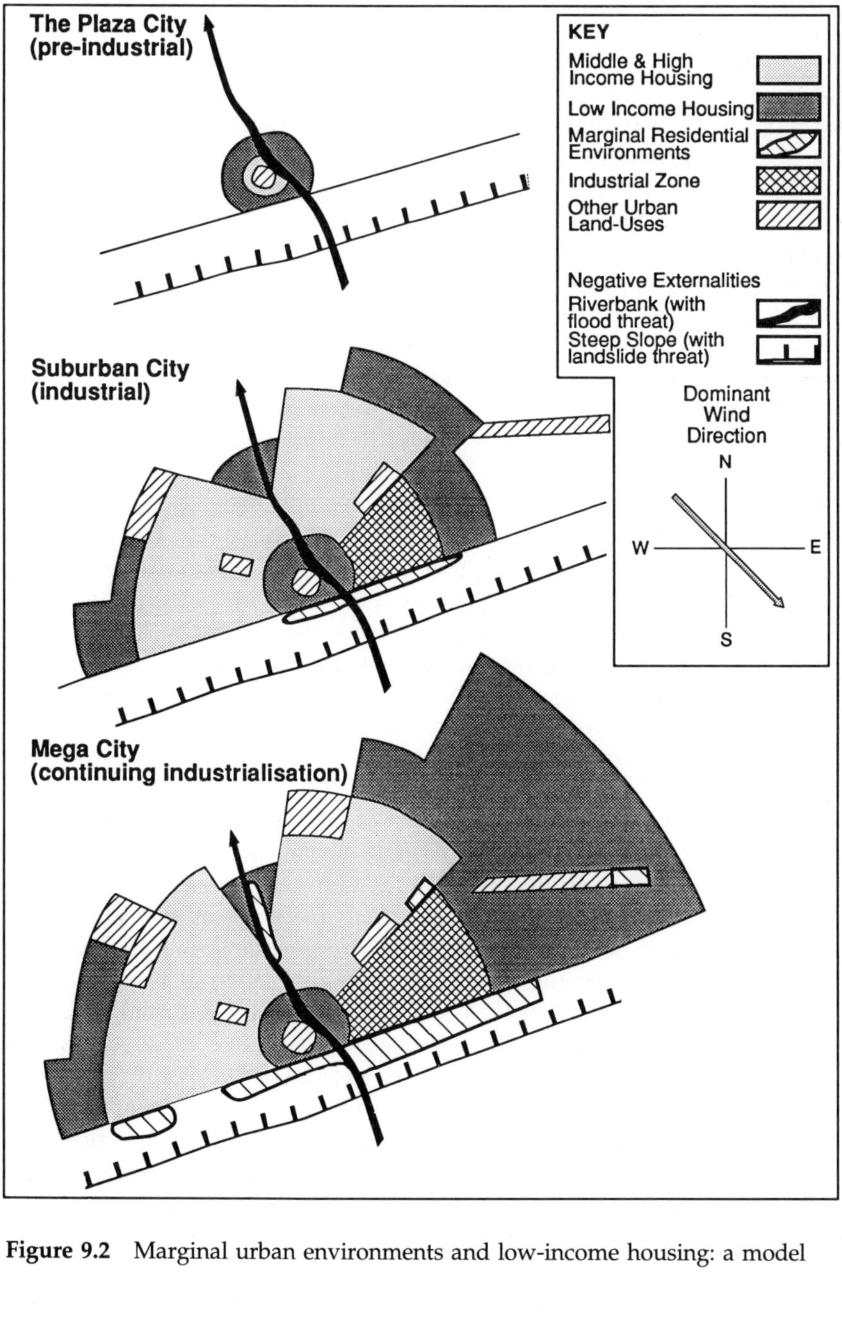

Figure 9.2 Marginal urban environments and low-income housing: a model

the industrial zone, are resolved by some of the peripheral poor through colonisation of the lower steep slopes.

The third stage of the model corresponds to the situation in some Latin American mega-cities today. Urban population has grown so large, and low-income neighbourhoods have become so constrained to the urban periphery through the expansion of middle-class suburbs and other legal developments, that despite widespread public transport daily commuting has become prohibitive for many in time and/or money costs. Accommodation costs are too high for them also in the central tenements, where very high residential densities yield lucrative profits for landlords and others investing in the land and housing markets. The poor and some of the not-so-poor are now largely squeezed out of existing accommodation possibilities, legal and illegal, other than marginal urban environments on the riverbanks and up the slopes. The problems and dangers of these sites are usually understood; but these are outweighed by their proximity to central and industrial workplaces, by the fact that they are cheap or even free of rent and their unattractiveness to formal land markets. Sites like these are havens for the urban poor, who are unlikely to be evicted from them because those with money and power and contacts have no use for them. Health problems are at their worst among those who inhabit the river banks downstream from the city centre; some of their shacks are washed away almost every year when the river is in spate, and many lives are lost in occasional floods. Those who inhabit the rocky slopes adjacent to the industrial zone suffer particularly badly from airborne pollution, though the shanty towns on the higher slopes towards the western periphery are relatively healthy places to live in; but the elderly find the climb hard where the roads cannot reach, and shacks are sometimes swept away by landslides.

In applying this model to African cities, the logic of our argument is as follows. The urban poor suffer marginal urban environments, disproportionate to urban populations as a whole, to a much greater extent in Latin America than in Africa. But this regional difference is not explained by intrinsically more awkward natural environments in Latin America: as Table 9.2 shows, it seems that major negative externalities are approximately as much of a feature of African cities as of Latin American and also Asian cities. Rather, explanation of the growing association between low-income housing and marginal urban environments lies in the progressive squeezing out of low-income neighbourhoods from land that is valuable for residential or other land-uses through the operation of increasingly developed land and housing markets in cities that are growing beyond a manageable size. Support for this explanation is yielded both by the evident rarity of environmental disasters in African cities compared with Latin American cities and by the apparently much lower bias towards the urban poor in the damage inflicted by African disasters.

If explanation indeed lies predominantly in the development of land and housing markets, and associated processes within overall urbanisation, two outcomes are noteworthy. First, the second stage of the model outlined above is more applicable than the third stage to African cities today; industrialisation, suburbanisation and the development of public transport networks began and expanded some decades later in Africa than in Latin America. Second, continuing urbanisation in Africa entailing development of land and housing markets along lines even approximately similar to those in Latin American cities will in coming years produce a significant increase in the African urban poor's occupation of marginal urban environments; and the probability of major urban environmental disasters will rise accordingly. Khartoum's flood disaster in 1988 might be indicative of an emerging trend of urban environmental calamities in African cities.

Marginal urban environments and the land issue

Thus far the discussion has focused on the nature, characteristics and spatio-temporal location of marginal urban environments, illustrating how these sites have become havens for low-income dwellers. From the discussion the environmentally marginal nature of these sites is apparent in terms both of their inherent and acquired characteristics (Gilbert and Ward, 1985). The former refers essentially to physical (natural) geographic features, while the latter refers to location in terms of services, transportation and, importantly in this context, exposure to a broad range of environmental problems.

Without significant changes in the attitudes and policies of governments to the housing problems of low-income groups, it would be expected that the future for most low-income dwellers will be one of constant insecurity. It would appear probable that low-income groups will locate increasingly on dangerous sites, either because these are the cheapest sites available or, alternatively, because they minimise the chance of eviction. Consequently, intra-urban segregation will become more marked with 'the poor in the worst-quality, worst-serviced and often worst-located sites with richer groups in the best-located, best-quality and best-serviced locations'. Furthermore, such processes will have a more profound impact in so far as 'cities' physical growth will be defined by where poorer groups can invade or illegally negotiate for land' (Hardoy and Satterthwaite, 1990, p. 235). Although the problem of marginal urban environments is very often discussed in technocratic terms (for example appropriate materials, structure and designs etc.), the canvas on which the problem of such environments is located is quintessentially political and institutional in nature. At the heart of these considerations is the relationship between a land policy and a low-cost housing policy. Specifically, it would seem apparent that the central issue is simply the failure of governments to provide an adequate supply of affordable, legal land sites for housing. As Angel *et al.* noted in 1983, 'land for housing the poor is gradually becoming an insurmountable obstacle in the housing struggle facing the growing cities of the Third World' (p. 3).

However, the 'land crisis' has not always been a characteristic feature of Third World cities. As the UNCHS (1987, p. 131) notes, given the pace of urban growth since 1950, urban land markets have been extraordinarily adaptive and 'what is surprising is not that they now face severe problems but that they function at all'. Thus at the end of the Second World War many cities were still quite 'porous' (Doebele, 1987) with large city-centre buildings encouraging subdivision, while the existence of large land lots allowed additional dwellings to be readily installed. Furthermore, well-located land owned by public and other institutions constituted informal 'land banks'. At greater distances from the city centre, settlements were established on land of low commercial value by invasion, quiet encroachment or by illegal subdivision.

However, for most of the major cities of the Third World the relatively easy availability of land for low-income dwellers has ended. In the central areas, saturation point has been reached, and in many instances expansion is only possible by an enlarged rental sector or by locating in the most marginal of environments. In peripheral areas the markets for both public and private land have tightened considerably. Over the years the attitudes and policies of governments towards squatting and invasion on public land moved from a relatively permissive position to one which viewed land as an irreplaceable asset which must be defended. This was reinforced by a planning environment which according to McAuslan (1985) focused on negative controls involving 'planning, zoning and any other measure which limits the use to which land can be put' (p. 64). At the present time land-use controls remain in place although they may be weak or unevenly applied, and as Mattingley (1993, p. 113) observes 'a number of market forces are working against government efforts to determine the uses of urban land'. In the private sector, with the lack of other secure investment opportunities, peripheral land has been the subject of speculative activity by high- and middle-income families, while construction companies who have prospered from urban growth buy land in the periphery to assure adequate sites for future projects.

As a consequence of the processes noted above the land market has tightened considerably for low-income dwellers. Nevertheless, as Hardoy and Satterthwaite (1989, p. 102) note, there is 'no lack of undeveloped or partially developed land on which low income housing sites could be developed'. In Sao Paulo, for example, it has been estimated that the extent of undeveloped land could accommodate a two-thirds' increase in population without any further expansion. However, as the above authors indicate, 'much of this land . . . is owned by individuals or institutions who keep it vacant because the growth of the city and the development of its economy is rapidly pushing up its value, despite the fact that no investment is made in developing it . . . Such land is rarely invaded; low income groups know that any attempt to do so would be crushed.'

Some, albeit limited, attempt to ease the land-supply problem for low-income housing has been associated with the sites-and-services projects which have been implemented in a number of developing countries. However successful these interventions are at a project level, the sites-and-services approach remains a negligible source of land supply for shelter for low-income groups. Attempts to scale up these pilot projects into continuous housing programmes have encountered major obstacles. Specifically, the process of land acquisition is a slow and expensive process (Zetter, 1984) and to avoid delays governments will tend to opt for less expensive low-value sites on the periphery which are easier to obtain. However, for the recipient of the plot this clearly is far from the ideal site-location, involving not only construction costs but also transport costs to and from work.

What, therefore, is to be done? In essence the solution is simple: what is required is a substantial increase in the supply of land available for low- and moderate-income families. However, as Table 9.3 illustrates, there exists a variety of approaches and instruments which could stimulate an increase in the supply of land of this type. Some writers recommend the use of positive government controls such as land-banking, land-pooling and compulsory acquisition (for example, McAuslan, 1985). However, there is a growing body of opinion which advocates the reduction of state intervention in the land and housing markets. In addition, a number of governments have stressed the importance of 'unblocking' the market as part of the solution to housing problems and lack of services. Certainly a number of important basic 'technical' actions exist which can help to 'unblock' a city's land market. For example, an adequate local cadastral and property registration system would appear to be essential to the lowering of urban land prices by increasing accessibility between buyers and sellers. Similarly, accurate inventories of publicly owned land are necessary for any effective policy concerning land and land supply. It is perhaps true that unblocking the market could increase the proportion of middle-income and lower-income households who can find legal housing solutions. Nevertheless, as Hardoy and Satterthwaite (1989) argue there will still remain a large proportion of individuals and households too poor to enter this market. For many of these households 'market solutions' are clearly not sufficient, and in some situations 'negotiated solutions' involving some concession to the households concerned would be appropriate. An example of this latter type of solution would be the land-sharing schemes as implemented in Bangkok (Angel and Boonyabancha, 1988).

Table 9.3 Increasing the supply of land for low-income housing: a typology

A.	INCREASING SUPPLY THROUGH DIRECT PUBLIC ACTIONS
1.	Accelerated construction of service systems and improvement of transportation.
2.	Improved methods of cost recovery (that is the unearned increment resulting from public investment).
3.	Reducing the attractiveness of urban land as a vehicle for the storage of capital (taxation, land-use controls, credit controls, price freezing).
4.	Prevention of unnecessary destruction of existing low-income settlements.
5.	The need to modernise cadastral and land-registration systems.
6.	Better legislation for compulsory acquisition.
7.	Inventories of all publicly owned land.
B.	INCREASING SUPPLY THROUGH IMPROVED INTERACTION OF PUBLIC AND PRIVATE (INCLUDING INFORMAL) SECTORS
1.	Supporting informal systems.
2.	Joint private-public development.
C.	MAKING EXISTING SUPPLY MORE EFFICIENT
1.	The need to re-examine the question of appropriate standards: — lot sizes — excessive building codes and standards for infrastructure and services.
2.	Community participation: mobilising the energies and ideas of local organisations.

Source: Adapted from Doebele, 1987; pp. 120–30.

The implication of the foregoing discussion is quite clearly that in many countries there is a pressing need for the reform of urban land policies and institutions (see Farvacque and McAuslan, 1992). In this context, the immediate issues requiring attention are readily identifiable and would include *inter alia,* the respective roles of public interventions, market mechanisms and traditional authorities; the balance between social and economic considerations; and the form of appropriate administrative and legal institutions. However, it would be naive to suggest that in such a complex arena as land and its management simple administrative and managerial solutions are sufficient. In addition to the proximate issues just noted, the reform of urban land policies is premised on a wider set of issues of governance, including participation and decentralisation in government, and transparency and accountability in adminstration. The consequences of failing to address this agenda will inevitably reinforce the tendency for many low-income dwellers to locate in the most marginal of environments, and one can only venture the pessimistic prognosis that the types of chronic problems and environmental disasters discussed earlier in this chapter will become increasingly common.

References

Abu Sin, M.E. and Harbi, M., 1991, 'Greater Khartoum's vulnerability to disaster hazards: the case of the 1988 rains and Nile flood', in M.E. Abu Sin and H.R.J. Davies (eds), *The Future of Sudan's Capital Region Khartoum*, Khartoum University Press.

Agarwal, A., 1990, 'Here's mud in your house', in *Third World Guide 91/92*, Montevideo, Instituto del Tercer Mundo, p. 35.

Ahmad, A.M., 1989, 'Housing submarkets for the urban poor: the case of Greater Khartoum', *Environment and Urbanization*, **1**(2), pp. 50–9.

Angel, S. *et al.* (eds), 1983, *Land for Housing the Poor*, Singapore, Select Books.

Angel, S. and Boonyabancha, S., 1988, 'Land-sharing as an alternative to eviction: the Bangkok experience', *Third World Planning Review*, **10**(2), pp. 107–8.

Baross, P., 1983, 'The articulation of land supply for popular settlements in Third World cities', in S. Angel *et al.* (eds), 1983, *Land for Housing the Poor*, Singapore, Select Books.

Baross, P. and Linden, J. van der, 1990, *The Transformation of Land Supply Systems in Third World Cities*, Aldershot, Avebury.

Clarke, J.I. *et al.* (eds), 1989, *Population and Disaster*, Oxford, Blackwell.

Claval, P., 1991, 'Specificities and problems of hillside cities', *Regional Development Dialogue*, **12**(2), pp. 14–28.

Doebele, W.A., 1987, 'Land policy', in Rowdin, L. (ed.), *Shelter, Settlement and Development*, Boston, Allen & Unwin, pp. 110–32.

Doornkamp, J.C., 1985, *The Earth Sciences and Planning in the Third World*, Liverpool, Liverpool University Press.

Douglas, I., 1983, *The Urban Environment*, London, Edward Arnold.

Dwyer, D.J., 1975, *People and Housing in Third World Cities*, London, Longman.

Environment and Urbanization, 1989a, 'Environmental problems in Third World cities', **1**(1).

—, 1989b, 'Beyond the stereotypes of slums; how the poor find accommodation in Third World cities **1** (2).

Farvacque, C. and McAuslan, P., 1992, *Reforming Urban Land Policies and Institutions in Developing Countries*, Urban Management Program, Policy Paper 5, Washington, World Bank.

Gilbert, A. and Ward, P., 1985, *Housing, the State and the Poor: Policy and Practice in Three Latin American cities*, Cambridge, Cambridge University Press.

Goethert, R., 1991, 'Lessons from squatters in Latin America: housing on hillsides for low income groups', *Regional Development Dialogue*, **12**(2), pp. 57–75.

Hance, W.A., 1970, *Population, Migration and Urbanisation in Africa*, New York, Columbia University Press.

Hardoy, J. and Satterthwaite, D., 1989, *Squatter Citizen: Life in the Urban Third World*, London, Earthscan.

Hardoy, J. and Satterthwaite, D., 1990, 'The future city', in S. Cairncross, J. Hardoy and D. Satterthwaite (eds), *The Poor Die Young: Housing and Health in Third World Cities*, London, Earthscan.

Jimenez, R.D. and Valasquez, A., 1989, 'Metropolitan Manila: a framework for its sustained development', *Environment and Urbanization*, **1**(1), pp. 51–8.

Jones, T., 1988, *Corporate Killing: Bhopals Will Happen*, London, Free Association Books.

King, A.D., 1991, 'History and culture in the development of the Himalayan hill station: the case of Shimla', *Regional Development Dialogue*, **12**(2), pp. 114–26.

McAuslan, P., 1985, *Urban Land and Shelter for the Poor*, London, Earthscan.

O'Connor, A., 1983, *The African City*, London, Hutchinson.

Silas, J., 1989, 'Marginal settlements in Surabaya', *Environment and Urbanization*, **1** (2).

Save the Children Fund, 1990, *Prospects for Africa's Children*, London, Hodder & Stoughton.

Susman, P., O'Keefe, P. and Wisner, B., 1983, 'Global disasters, a radical interpretation', in K. Hewitt (ed.), *Interpretations of Calamity: From the Viewpoint of Human Ecology*, London, Allen & Unwin, pp. 263–83.

UNCHS (HABITAT), 1987, *Global Report on Human Settlements 1986*, Oxford, Oxford University Press.

van Lindert, P., 1991, 'Moving up or staying down? Migrant-native differential mobility in La Paz', *Urban Studies*, **28**(3), pp. 433–63.

Ward, P., 1990, *Mexico City: The Production and Reproduction of an Urban Environment*, London, Belhaven.

Weir, D., 1986, *The Bhopal Syndrome – Pesticide Manufacturing and the Third World*, London, Earthscan.

Zetter, R., 1984, 'Land issues in low-income housing', in G.K. Payne (ed.), *Low Income Housing in the Developing World, The Role of Sites and Services and Settlement Upgrading*, Chichester, John Wiley, pp. 221–32.

1 Agadir; Plateau seen from fishing port, 1978. The Plateau was the centre of Agadir until the 1960 earthquake, which had its epicentre here. Urban development has been strictly planned since then on the lower ground away from the Plateau, and all that remains here is the ruin of the kasbah amid open space. (*Photograph*: Martin Metcalfe, University of Durham)

2 Mexico City; earthquake-damaged medium-rise buildings, 1985. The November 1985 earthquake off Mexico's Pacific coast wrought more damage in Mexico City than anywhere else. The buildings worst affected, unusually, were medium- and high-rise structures like these on the soft alluvial sub-soils. The most affected social group were the lower-middle classes who reside in such buildings in greater numbers than other social groups. (*Photograph*: Martin Degg, Chester College)

3 Caracas; hillside and squatter dwellings after landslide, 1986. The areas that suffer most from the frequent landslides in and around Caracas are squatter neighbourhoods which have been constructed on marginal terrain, from which they are less likely to be evicted than from more desirable land. This September 1986 landslide in the suburb of La Vega destroyed dozens of squatter dwellings constructed on the steep slopes of the hill. (*Photograph*: Virginia Jiménez-Díaz, Universidad Central de Venezuela, Caracas)

4 Delhi; borrow-pit and site of previous landslide, 1993. Borrow-pits, from which earth has been taken for use as building material, are evident around cities in many Third World regions. They may serve several purposes, but have to be used with caution. A landslide on the far side of this borrow-pit, on the edge of the Delhi squatter neighbourhood of Samaipur-Badali, killed eleven people in 1990, since when construction has been set further back. The borrow-pit is used today for growing vegetables and as a latrine for the women of the neighbourhood. (*Photograph*: Hamish Main, Staffordshire University)

5 Cairo; rooftop dwellings and haze created by atmospheric pollution, 1993. With a population of over 15 million, Cairo faces a range of housing and environmental problems. There are now over half a million rooftop dwellers in a variety of accommodation; those shown here are in the suburb of Mohandiseen. Winds from the north bring poisonous pollutants from lead and zinc smelters, while from the south the city is polluted by emissions from steel and cement factories; and dust blows in from the surrounding desert. (*Photograph*: Stephen Wyn Williams, Staffordshire University)

6 Hong Kong; privately built high-rise accommodation in Kowloon's walled city, 1992. As a consequence of its disputed sovereignty between China and Britain, Hong Kong's health regulations and licensing rules have never applied in Kowloon's walled city. Containing hundreds of unregulated manufacturing and service activities, it also provided a focus for child prostitution and drug rings. Scheduled for demolition by mid-1994, 28,000 residents and 900 commercial undertakings were cleared by July 1992. (*Photograph*: Stephen Wyn Williams, Staffordshire University)

7 Katsina; domestic architecture and neighbourhood space, 1977. A long tradition of domestic architecture using sun-dried clay is still evident in the savanna and sahel regions of West Africa. The use of local clay in building is inexpensive and allows flexibility in construction; it produces a visual harmony between natural and built environments, and a cultural continuity between urban and rural areas. Beyond this finely decorated compound entrance in Mani, Katsina, lies a series of rooms and courtyards that offer shade and privacy. (*Photograph*: Michael Mutter, University of Westminster)

8 Kano; mixed construction types in Dala neighbourhood, 1989. In Nigerian cities like Kano, a mix of construction types is apparent in many neighbourhoods; clay walls and roofs are still maintained and built to traditional styles, but concrete blocks and metal are increasingly used for walls and roofs. Borrow-pits like that in the foreground fill with rainwater during the rainy season, attracting mosquitoes; piles of timber and fuelwood for local consumption have been transported from hundreds of kilometres away, while trees are valued especially in the older neighbourhoods like this one (Dala) within the ancient city walls. (*Photograph*: Hamish Main, Staffordshire University)

Section 3

Housing, environmental problems and resources

10 Urban housing and services in Anglophone West Africa: coping with an inadequate environment

Margaret Peil

Most of the people living in African cities are poor, and urban conditions and amenities reflect their own and their country's poverty. While there are a few 'squatter settlements', 'slums' or other places known locally to accommodate the very poor, most areas have a mixture of incomes and standards of living; quite a few people living in low-income areas (especially businessmen but also civil servants and professionals) are not poor by local standards. Thus, environmental problems which might affect only the poor in other parts of the world are part of the daily lives of most West African urban residents. Where they have a choice, many people choose to live in less well-serviced areas for family reasons, because they cannot find alternative housing in a rapidly growing city or because they choose to spend their money on other things (such as buying land at home or building up a business) rather than on rent. People frequently move within urban areas because the houseowner wants the room, to be nearer to their jobs, for lower rent or to escape unfriendly neighbours or the police. In addition, outside major cities there is little environmental differentiation between one area and another; residents share common problems even where a few areas can be considered at the bottom environmentally.

Indigenes of the town may be concentrated in one or more areas or scattered, as houseowners throughout the urban area. They see the town's land as theirs, and have a relatively strong motivation to improve their environment or at least maintain it. Migrants, on the other hand, are seldom committed to the town or the area they live in, until they decide to settle down and build a house. This is especially true if they are ethnically different from the owners of the town. If they intend to go home eventually, why worry about the urban environment? Residence on a housing estate, a relatively new phenomenon for the poor of Anglophone West Africa, can engender a desire for improvement of the services or environment, since the establishment of estate boundaries puts residents into a group which can make demands on the government. For example, the president of the Isolo Low Cost Housing Estate Landlords' Association complained of high rates; lack of primary schools, water and electricity, post office and telephones; inadequate roads and markets; and crime on the estate (Peil, 1991, p. 160). However, the really poor seldom get 'low-cost' estate housing and are less able to complain about the conditions in which they live. People usually need some security before they can join together to make effective political demands.

This chapter examines the housing situation as related to the urban environment. West African urban housing is usually legally recognised, overcrowded and underserviced. I begin with tenure, because this affects people's attitude both to their housing and to the town in which they live.

Tenure

The majority of urban residents in West Africa are tenants with no interest in ownership. They find a room at a price they can afford or find a relative or friend with whom to share the rent. Most houseowners rent out some rooms to increase their income; single-family housing is rare and limited to the elites or large, extended indigenous families. There is relatively little squatting and what there is depends more on bureaucratic definitions than on an absence of land-tenure arrangements between the original owner and the current user (Peil, 1976; see also Konadu-Agyemang, 1991). Most West African migrants have not seriously considered the possibility of owning urban land because they prefer to return to their villages and are urged by their families to spend any savings on land and housing at home; they cannot afford the costs even though these were relatively modest until recent years; and/or because rents have been so low that the commitment and risks of urban property development have seemed unnecessary. This may be changing since the greater likelihood of surviving into old age and the presence of their children in town have made urban retirement preferable for an increasing number of people. This is especially true of women, but women seldom have the resources to build a house. Few people have a pension, and inflation makes these grossly inadequate. The money to be made from renting out rooms provides security for older men and widows who have inherited a house. Thus, while rural housing is for family, urban housing is for investment – if funds are available to buy land and build.

In the past, the willingness of investors to build tenements for low-cost rental by the room has meant that housing demand and supply have been closer in Anglophone West African countries than in many other parts of Africa, and squatting (as commonly defined: using land without the owner's permission) has, by and large, not been necessary. Rents have generally risen more slowly than inflation and thus have not been a stimulus to home ownership. The Nigerian urban consumer price index shows that accommodation rose only half as fast between 1976 and 1989 as the total index and only 42 per cent as much as food (FOS, 1989). Other important factors have been owners' vigilance to ensure that family-owned land remains under their control and governments' reluctance to get involved in the provision of low-income housing. Poor people do not expect the government to provide housing for them, and it is difficult in densely populated cities to find land on which to squat successfully.

The preference for rental housing may change when a return to the rural home becomes problematic. Elderly people settle in small towns of Sierra Leone because they see life there as better than rural life (more services and security). Tanzanian migrants settle in Dar es Salaam because collectivisation of farm land means they have nothing to return to, and Zimbabweans may settle permanently in towns even though they prefer rural life because land for farming in the Communal Areas is too limited to provide an economic base for retirement. In all of these cases, house ownership in town has become more important, and pressure on urban land and the environment have increased.

Another factor is the rising cost of building as governments introduce more regulations. For instance, Sierra Leone requires all houses to be of cement, which is unenforceable in villages, but raises the costs of urban houses which might initially be built more cheaply. As the costs of both land and building materials are pushed up by speculation and inflation, the ability of the poor to become owners declines. When a small house could be built on the outskirts of a town for the equivalent of a year's income (which might be borrowed from kin or saved from business or an unexpected windfall), poor people could get a start in housing. These houses may provide minimum shelter when originally built, but are improved and incorporated into the town over time.

Many West African houseowners have built a room at a time, as they could accumulate the necessary capital; in recent years, there is more emphasis on building the whole house before moving in, possibly delaying occupation for some years unless the builder is rich. Since the prices of land and building materials have risen rapidly in recent years, the poor cannot participate at all. Rich people have or can borrow the capital, or they can charge potential tenants a substantial sum in key money. Thus, the house can be paid for in a year or two, providing more income for another venture. But higher costs require higher rents for an adequate return, and building for the poor has declined; the supply of low-income housing has fallen well behind the demand. In the 1970s a 'revolutionary' Ghanaian government strictly limited rents on the grounds of social justice. With no profits in sight, landowners let their houses fall to ruin or refused to rent out rooms, and almost no low-income housing was built. The severe shortage of housing which resulted eventually forced the government to relent (Tipple, 1982).

Land acquisition is often a long and hazardous business. Land for industrial estates has often been taken over by the government, moving indigenes and settlers to new areas as at Tema, but most urban land has remained in private ownership. In long-established towns, the land has been held by the extended families of the original settlers over many generations. As these towns have grown, some of this land has been sold to individuals wanting to build for their families and to speculators. However, agreement of all the co-owners is required for land to be sold, so considerable negotiation is needed and land rights are affected by family conflicts over personal influence and leadership (Aronson, 1978a). Sometimes family members sell land without getting permission from the family and unscrupulous entrepreneurs sell land that they do not own. In these cases, the unwary buyer may have to pay again. Landlords form neighbourhood associations to provide mutual support in case their rights are challenged (Barnes, 1986). These also help to control squatting, since non-resident landlords can be warned by their colleagues that their rights are in danger. In recent years, owners have been less willing to sell land. Large-scale owners in Lagos now prefer leasing their land, so they retain control in a rapidly rising market.

The lack of paperwork in many land sales has made large numbers of lawyers wealthy. The parties concerned merely shook hands in the presence of witnesses on payment of drinks or a small sum in cash. When there were papers, they were not in legal terminology, and given the local climate and housing conditions they were often lost or destroyed by insects. Boundaries based on trees or streams are subject to change over time. As a result, there have been many disputes as to who owns what land, and many attempts by enterprising developers to usurp the property of others.

Until recently, registration of ownership and planning permission have not been part of the land acquisition process in these countries, and most governments still have little control over land use. There are regulations about the proportion of the plot which can be built on and some building regulations, but enforcement is lax. Even newly developed areas mix housing and small business.

Since the 1978 Land Use Decree, urban land in Nigeria is supposed to be officially registered. The certificating authority requires proof of boundaries, which should eliminate land disputes, and development within two years, though this provision is seldom enforced. Owners can cover themselves by subleasing their property to what may appear to be squatters but are actually families paying relatively high rents to live in severely substandard housing. This serves to keep real squatters at bay, since there is no vacant land for them to use. For example, large landowners in the Olaleye Village area of Lagos use this method to hold land speculatively until it appreciates and can be sold at a large profit. Plots are rented by the month, year or five years. Given such short tenancy, only the most basic accommodation is built. To help pay the rent, part of the space in these shanties of wood or zinc sheets may be rented out to sub-tenants (Aina, 1990). Thus, land which in developed countries might remain empty for several years is as crowded with accommodation as land whose future is settled, but there is little pressure on the government to service the site.

Security for land/house owners is somewhat different than security for tenants, so they will be treated separately, beginning with owners. Legal possession provides little security if powerful people, public or private, want your land. Lagos provides several examples of government and private action which has severely disadvantaged low-income owners, but its relative rarity means that the evidence must be used with caution. The most famous is the Lagos Slum Clearance Scheme of the 1950s which aimed to remove low-income indigenous families from owner-occupied houses on Lagos Island (the centre of the city) to government-owned rental housing in Surulere, a suburban housing estate. This was in the build-up to independence, when the Nigerian government was eager to present a 'modern' capital city to the outside world. In official rhetoric, families could move back to their redeveloped land in the centre, but the impossibility of this was soon evident to owners. Families sharing compensation among a large number of kin would never be able to collect the money to build a new family house. In addition, fewer plots would be available, since considerable land was appropriated for commercial and government use.

Many of the indigenes forced from their homes moved within central Lagos or to low-cost suburbs where they could own houses with more space than the Surulere houses provided. Moving to Surulere involved considerable loss to traders who relied on large-scale traffic through local markets and streets, and family/neighbourhood ties were severely disrupted. The majority of residents of the Surulere estate today are not indigenes and have above average incomes.

The project was stopped by strong popular opposition, and Lagos Island still provides considerable low-income housing, in extremely crowded conditions, just behind the banks and office towers of the central business district. This area also accommodates relatively affluent single people and small families who find the location convenient. By popular demand, most Surulere houses have been sold to the tenants (Peil, 1991, pp. 167–70). The government would probably like to continue the clearance, but even military governments have found it politically impossible to do so. The lesson was well learned; when the Ghanaian government organised a survey in central Accra in the 1960s, interviewers were stoned and the project called off. Indigenes in this area were determined not to lose their houses; this is one of the most densely populated areas in Accra.

'Strangers' are easier to dispose of. A more recent case of government action concerns Maroko, which in the 1980s probably had the lowest average income in Lagos. Maroko was initially a fishing community on swampy land at the end of the Lekki Peninsula. The Lagos Executive Development Board, the current planning agency, moved people there in 1958 so that Victoria Island and Ikoyi could be drained and developed for Europeans and wealthy Africans. By 1962 there were twenty-two 'villages'/neighbourhoods in Maroko; local chiefs supervised land distribution in the traditional way. Migrants arrived in large numbers as housing was poor but cheap. There was no safe water supply, no drainage and irregular electricity, but it was convenient for work in the city and the government provided a school and a clinic. However, by the early 1980s government planners wanted to develop the Lekki Peninsula, including Maroko, to expand the nearby high-income area. This is a good example of the premise that land originally considered environmentally marginal can be upgraded if important people like the location.

Though they had acquired their land through procedures which were legal at the time, owners were called squatters when the government wanted to clear Maroko. The state government demolished one neighbourhood in 1984, but a change of governor provided some reprieve. However, in 1990 the whole settlement was bulldozed, making about 300,000 people homeless. Until the land is actually developed by and for the elites, one must sympathise with people who squat on their former land. Promises to provide new land have come to nothing, and finding cheap accommodation in an extremely crowded Lagos is very difficult. Eight months later, residents were demanding 500,000 Naira from the government for each owner and 100,000 Naira (about £15,000) for each tenant household, for a total of 85 billion Naira (*West Africa*, 1990, p. 2166; 1991, p. 1251). They won't get much, but lawyers may impress the government with the idea that they cannot take people's land and houses without compensation.

There have been other incidents. Many people in both Lagos and Abeokuta have found to their cost that governments determined to create or widen roads can destroy their houses with impunity. One Lagos businessman who went to court when his fences and later his house (built with official permission at the requisite distance from the road) were destroyed was told by the judge that, as a northerner, he should not be in Lagos at all. In Ghana, soldiers have ignored official plans and taken land for their own houses in spite of legal use by others. Owners who lack powerful contacts are likely to lose any legal battles. Since numbers can represent power, especially when an election is imminent, it is important for landowners to join together to protect their interests. This is what happened in central Lagos, halting the destruction of the redevelopment scheme and saving many houses.

Insecurity poses different problems for tenants, for whom rent often costs over a quarter of their income. There is considerable mobility to be near job, family or friends, but conflict with owners and fellow tenants is also an important factor. Although rents generally rise more slowly than inflation and tenants can technically go to a rent tribunal to protest exorbitant rises, rents can legitimately be raised whenever a new tenant comes in, and written leases are rare. Owners have various ways of getting rid of tenants who either cause trouble or pay less than the going rate for their accommodation. Rooms can be demanded for 'redecorating' or because they are wanted for family or relatives (always a valid possibility, though kin do not pay rent). Where housing is in short supply, moving may involve paying a fairly high sum as 'key money' or sharing with already crowded relatives or friends until a room can be found. It is the very poor (the elderly, widows, the unemployed) who have most to fear. In the past, they could often develop relationships with, and provide services for, a resident landlord and thus maintain their place. However, in recent years there has been an increase in speculative housing owned by non-resident landowners and administered by an agency, especially in major cities. This business arrangement is much less hospitable to the needy.

Housing density and services

Rapid growth and high densities in African cities are a major environmental problem since in addition to the stress they cause residents, they make it difficult to provide services and keep order. Once accommodation has been found, urban residents must often cope with its inadequacies in space, condition and services. A large proportion of households are crowded into a single room in a building which has many other tenants, is not regularly maintained and where facilities and services are a long way from what is taken for granted in European houses in the 1990s.

Urban densities are very high in West Africa, though there are few houses of more than two storeys except in the largest cities. In 1988, the estimated density for the Lagos Metropolitan Area was 20,000 per square kilometre, up from 13,665 in 1974 (Lagos State Government, 1989, Table 1). Newer tenements in many parts of the city tend to have four storeys of twelve rooms per floor (the legal limit), with as many as 150 people in a building, plus a shifting population of visitors staying with relatives.

Smaller towns, with less demand, have smaller houses and are somewhat less crowded, though densities are still well above those of low-income areas of the United Kingdom (see Table 10.1). A study of eight towns in Ghana, The Gambia and Nigeria found an average of six to nine rooms in one- to two-storey houses, with between 1.4 and 3 people per room (Peil, 1981, p. 125). The Nigerian General Household Survey found an average of 4.4 persons per household; rural households averaged 5.18 (FOS, 1985a, Table 1.1).

Suburban densities tend to increase over time, as the city grows around them. As an example, the Ajegunle area of Lagos had about 90,000 people in about eight square kilometres of swampy land in 1972. By 1990, the built-up area was only slightly larger, but the population had grown to about 500,000. Residential densities in Lagos in 1977 varied from 100 people per hectare in Ikoyi to 607 in Agege on the outskirts, 1,200 in Mushin and 2,150 on the indigenous north side of Lagos Island, where there were an average of 5.4 people per room (*Master Plan*, 1978; 1980, vol. 1, p. 115). Densities have probably increased since then, as Lagos's population is now nearly three times as large. In Accra, the average of 1–1.5 people per room in elite areas can be compared to 4.0 in Achimota, 3.6 in the indigenous core and 3.5 in Sabon Zongo – all mainly low-income areas (Ghana Government, 1990, pp. 61–4). Thus averages for a city hide the extreme crowding in which many people live.

Table 10.1 shows an average of 2.2–2.9 persons per room in many West African cities, with the notable exceptions of The Gambia (small towns in a very small country), Kumasi (the centre of the cocoa industry, with very

Table 10.1 Housing density in some Anglophone West African cities

Location	Year	Mean number of persons per room	Percentage of households in one room	Source
Gambia:				
Banjul	1976	1.9	27	Peil, 1981, p. 125
Serekunda	1976	1.8	21	Peil, 1981, p. 125
Ghana:				
Accra	1989	2.9	46	Ghana, 1990, p. 61
Ashaiman	1989	3.1	51	Ghana, 1990, p. 64
Kumasi	1980	3.6	70	Tipple & Hellen, 1986
Tema 4,6,9	1989	2.3	13	Ghana, 1990, pp. 62–3
Nigeria:				
Benin City	1974	2.2	48	Sule, 1978, p. 70
Lagos	1985	3.5	95	FOS, 1985a, pp. 37, 54
Uyo	1979	2.9	44	Animashaun, 1981, p. 24
Warri	1974	2.6	60	Sule, 1978, p. 70
Sierra Leone:				
Bo	1981	2.5	45	Peil (survey data)
Freetown	1981	2.5	46	Peil (survey data)
Kenema	1981	2.4	44	Peil (survey data)

little recent building) and Lagos (a primate city where housing is very expensive). The rate of growth and local economic conditions are more important than the size of city in increasing the demand for housing. Banjul was growing slowly; Serekunda was growing more rapidly, but land was available so people could build or find a room to rent. Accra is considerably larger than the other Ghanaian cities, but Kumasi appears to be more crowded. Tema was growing rapidly, but middle-income housing (Communities 6 and 9 in this survey; Community 4 is for low-income households) is significantly less crowded than low-income housing because most households have more than one room. Tema is a government-built town where most accommodation is allocated by employers; its standards, services and convenience to the workplace mean that people prefer it to cheaper but less adequate and more distant housing in Ashaiman. When more jobs are available, some Tema rooms are sublet and the density increases.

Freetown had the same or lower levels of crowding than other Sierra Leonean cities, though the large number of refugees who have fled there in the last two years have certainly made it more crowded. A larger mean household usually also means more crowding because so many families live in a single room; children and relatives are especially related to crowding in the West African situation. Warri demonstrates that a large proportion of single-room households need not mean severe overcrowding if the households concerned remain small. In the Warri case, many were young workers who had not yet built up large households.

The proportion of households in one room is related to the proportion of industrial workers in the population and hence attractiveness to young migrants and the rate of growth. With the exception of Tema, towns and neighbourhoods near industrial estates tend to have high proportions in single rooms, whereas less industrialised towns have considerably fewer single-room households because families have settled there over a long period. Very large households, with over thirty members, have been more common in Sierra Leone than in Nigeria, where the urban population tends to be less stable.

High densities have many implications for the physical and social lives of inhabitants; services are often grossly inadequate, especially in newly urbanised areas (see Table 10.2). Although the majority of urban households have electricity, the service is often erratic. It has improved in recent years in Nigeria but deteriorated in Sierra Leone as the country approached bankruptcy. Towns and cities in southern Ghana are generally well served by electricity from the Volta Dam; the grid has recently been expanded to the north.

None of these countries is able to provide the minimum water supply considered necessary to its urban residents, though several large projects have improved the quantity and quality of water supplies in recent years. By the mid-1980s, about two-thirds of Nigerian and Ghanaian urban

Table 10.2 Housing services and amenities of households in some Anglophone West African cities (percentages of total in survey)

	Nigeria urban	Lagos	Makurdi	Accra	Kumasi
Electricity	73	93	69	91	na
Pipe-borne water	66	47	50	72	67
Rubbish collection	16	38	2	na	na
Piles in neighbourhood	41	25	53	33	na
Kitchen:					
None	10	1	na	44	na
Shared	63	86	na	28	na
Exclusive	22	13	na	27	na
Toilet:					
None	21	<0.5	2	29	22
Shared	57	86	95	50	most
Exclusive	22	14	3	20	few
Bathroom:					
None	8	0	na	5	na
Shared	67	89	85	73	na
Exclusive	24	11	na	22	na

na = not available

Sources: Stren, 1985, pp. 25–8; FOS, 1985a, Tables 12, 13, 20–2; Tipple & Hellen, 1986; Ghana, 1990, pp. 84–8.

households had pipe-borne water, mostly treated. However, there is considerable variability in what is available. Many metropolitan households in new areas are not served, and only 14 per cent of urban households in Bauchi State, in northern Nigeria, have piped water. The installation of a pipe-borne water supply still leaves many problems, since this water is easily polluted through leaking pipes or inadequately protected tanks and the supply is often grossly inadequate – the water may flow for only an hour or two each day. Households without their own supply mostly get their water from public standpipes (often not in working order), private wells or streams; no figures are given of households which buy their water by the bucket or tank load – a regular business in some areas. Wells and streams also become contaminated when they are too close to toilet facilities, and water bought by the bucket or stored in containers in the compound should be boiled before drinking – an expense which most people cannot afford. Interviewees in both high- and low-density areas of Makurdi listed a poor water supply as the main neighbourhood problem (Stren, 1985, p. 28). Rubbish collection is a perennial problem which will be discussed below under environment.

A majority of urban houses contain several households which share the kitchen, bathroom and toilet facilities. Except in elite households, kitchens usually have little equipment and may also be used for sleeping. Partly because the kitchens are small, insecure and overcrowded, a large number of people ignore them and cook on their verandahs or in the hallways, especially in Lagos. In between meals, cooking utensils and food are kept in their already overcrowded rooms. This is also true for the 10 per cent of the households with no kitchen.

Cooking is normally with kerosene or wood, since only the elites can afford to cook with gas or electricity and supplies of these are often unreliable. An inexpensive gas cooker now being introduced in Accra may expand the use of gas if it becomes widely available. Kerosene has been cheap in Nigeria but wood is difficult to get and use, especially in large southern cities. When a shortage of kerosene occurs, long queues develop, since users often have no alternative. Wood is mostly used in the north, where villagers used to bring it into the cities by donkey, returning with nightsoil to fertilise their fields; today it is mostly transported by lorry. Firewood is in increasingly short supply, and urban residents need cheap alternatives. Charcoal is used in many towns, but producing this is destroying trees faster than they can be replaced (Cline-Cole, 1987; 1989).

In 1984–5, about a fifth of Nigerian urban households had no toilet facilities at all; this proportion was over two-fifths in Kwara State and the highly urbanised Ogun State, where many small and medium-sized towns are over a century old. Even where toilets are available, they are often inadequate for the number of people who use them. There are often only one or two toilets (pit, pail or flush) for the whole compound (one for twenty-five or more people), though the owner's family may have one to themselves. Flush latrines exist only in high-cost areas and can only function when the water supply is adequate. Pail or bucket latrines are still common, in spite of the health hazard they pose, because they do not rely on running water. Few labourers will take the emptying jobs, and shortages of vehicles and places to dump the faeces mean that streams (an alternative source of water for households) and residential areas near the official dump are contaminated. Many areas have public latrines, but these have too large a catchment area and seldom get the regular service they need. Most are extremely unpleasant for those who must use them. The alternative is to use 'the bush' – vacant lots or other open spaces – another source of contamination, since these also serve as children's playgrounds.

Ghanaians are making increased use of KVIP (Kumasi ventilated improved privy) toilets for schools and estates in low-income areas. These are similar to the Blair toilets used in Zimbabwe except that they are built of more permanent materials in blocks of four. Ideally, composting takes place on site and there is no smell. However, these toilets can only be used with relatively low densities and are difficult to keep clean when used by more than one household, and most Ghanaians taboo the use of composted excrement (Tipple and Hellen, 1986). In northern Ghana, some people have a cultural objection to using any small room used by several others as a toilet; out of doors is preferred, even in cities.

Bathing facilities are easier to provide than either kitchens or toilets, though they are usually shared. As in villages, most houses have an enclosed area in the compound to which the bather brings a bucket of water. The main problem is drainage. Public bathrooms exist in some parts of Accra for those whose houses lack even this basic amenity.

Houses also vary considerably in quality. Less than half of the houses in Nigerian towns in the mid-1980s were built of cement blocks. Though the proportion was over 90 per cent in Lagos and, rather surprisingly, underdeveloped Gongola State, it was less than 7 per cent in Kano State, where nearly 90 per cent of houses still had mud walls. These are often given a coat of cement and are painted, which makes them last much longer. Walls of corrugated iron, planks or thatch are rare except in squatter areas. The majority of dwellings have cement floors, but most still have corrugated aluminium or iron roofs, giving whole cities a rusty look (FOS, 1985b, Tables 2.14–2.17). High-income houses in Accra tend to have asbestos/slate roofs, in spite of the known hazards of asbestos. Most houses are of cement blocks, though landcrete and sandcrete are low-cost alternatives in both high- and low-income areas. Nearly a fifth of Accra houses were built before 1940 and only 36 per cent since 1970. Given intense use, many houses are deteriorating, though few have major cracks in the walls (Ghana, 1990, pp. 79–82). On the whole, housing provides a place to sleep and protect one's goods; it is not looked to for comfort.

Environment

A population living at high densities with inadequate services tends to take environmental problems for granted; there is little they can do about them. On the whole, the environment gets relatively little attention from either politicians or the public; both are faced with greater problems. Most of the population has come from villages without electricity, pipe-borne water or sanitary services and is struggling for survival; their expectations of urban services are not high. There appears to be less environmental differentiation between low- and high-income neighbourhoods in West Africa than in South American or European cities. The level of service provision may be better on high-income estates, but in so far as these are relatively new the roads may be unmade, the water and electricity supplies irregular, rubbish collection sporadic and other services inadequate. For example, much of Lagos is low-lying and swampy, and high-income Victoria Island suffers from the same flooding as low-income Ajegunle. Areas settled by indigenes may have better services than those settled by strangers because there has been more time for development and because indigenes have greater political power, but if the strangers hold important government positions their neighbourhoods may get improvements not yet available to the rest of the town.

Urban roads are a major problem; in unplanned new areas they may be nonexistent. Streets of major cities are usually too narrow for traffic and in need of repair. They are crowded with a variety of public and private vehicles and pedestrians. Bicycles and motorcycles are popular in some towns but not in others. There are many pedestrians in all towns and cities, both because riding costs money and because it may be slower than going on foot when the roads are jammed. Major employers and some private schools run their own buses because the public service is inadequate. Where they are available, buses and minibuses are extremely crowded and taxis frequently serve several customers at once. There is seldom any place to pull off the road to take in or let out passengers, a situation which often slows traffic considerably. Wagon pullers also slow traffic in many areas, and cattle may be herded down the street in the midst of traffic. Where rubbish is thrown into the street, especially in low-income areas where there is no public rubbish collection, this interferes with traffic as well as being a health hazard (Inyang, 1978).

Private cars create massive traffic jams as the elites cross cities delivering their children to school and reaching their workplaces. Lagos's main contribution to traffic control has been to limit private cars on weekdays: either Monday, Wednesday and Friday or Tuesday and Thursday according to the first number on the licence plate. This has cut down the numbers somewhat, but the wealthy buy more cars; widening roads and providing overpasses has done at least as much to speed up traffic.

Torrential downpours compound the traffic problem, as streets may be flooded for days. The drainage ditches along the roads help to alleviate flooding, at the cost of smells and disease from uncleared rubbish and danger to the unwary pedestrian or child. Paradoxically, Lagos may have too much water and too little (for drinking) at the same time. The water supply is only about a quarter of the need estimated in the *Master Plan* (1980, 536). Water pollution of the lagoon and shallow wells is endemic, and many people must buy water because their area has no public water supply.

Air pollution is increasing over major cities but has not attracted much government concern. For example, the discharge from factory chimneys in Lagos combines with traffic fumes and industrial and home fires to produce considerable smog which light winds do little to disperse. Noise pollution is an even worse problem, increasing with relative affluence; blasting radios and record players from homes and shops and amplified Muslim calls to prayer supplement or encourage high-volume conversation. Abeokuta has 2,500 taxis, which hoot from early morning to late evening. Traffic discipline seems to have improved in many towns in recent years, but horns are still a vital part of the scene.

While African households do not discard the large amounts of packaging which European households produce, food preparation and informal sector employment produce large quantities of solid waste, which local authorities are ill-equipped to dispose of. Rubbish thrown along the streets, especially around markets, can block roads as well as be a health hazard. Only a small proportion of Nigerian urban household rubbish is collected by the government or private agencies (12 per cent overall, but 62 per cent in Imo State and 32 per cent in Sokoto State in 1984–5). The Lagos Waste Disposal Board collected 2.4 million metric tonnes of rubbish in 1988. It needs to collect several times that, but disposal remains a problem. A huge sanitation plant has waited more than a decade to be commissioned. Even if much is still thrown into the lagoon, the city is cleaner than it used to be. Some rubbish is burned, buried in the compound or disposed of in the bush, but considerable quantities are dumped in gutters or piled in the street – a potential source of malaria, cholera and other diseases and of water overflow when it rains. A few towns have established neighbourhood collection points, but unless these are emptied regularly they cause the same problems as irregular dumping. There has been a decline in the use of unauthorised heaps in Nigeria through government publicity and penalties; the proportion of rubbish disposed of within the compound has more than doubled, since waste collection is still not available in many areas (FOS, 1985b, Table 2.21).

Environmental sanitation is more problematic in smaller towns, since they often lack public latrines, and lorries for rubbish collection cost more to buy and maintain than many local councils can afford. There may be no vehicular access to some high-density areas, notably in older cities such as Ibadan (Pasteur, 1979; Sule, 1981). Aba, a commercial centre in south-eastern Nigeria, had many environmental problems after the civil war, but hometown enthusiasm helped to improve local conditions. Initially, they put out a few large rubbish containers. These were well used but not large enough for the job; each location soon had mounds of rubbish which would not fit into the container. By 1978, Aba had numerous small containers along major streets, emptied every other day, and the town was a model of cleanliness, even in the vicinity of the market.

The Nigerian government has established Environmental Sanitation Day, on the last Saturday of every month, to make people more aware of their environment. This was the successor to a War against Indiscipline and a War against Filth. Everyone, rich and poor, must stay at home and clean their compounds and the surrounding area between 7 and 10 a.m. Piles of rubbish are left out for government collection, and people are made more conscious of keeping things tidy. While piles of rubbish exist in Accra and other Ghanaian towns, they are usually small, so such drastic action is not required.

Several towns in southwestern Nigeria have a different approach to improving the environment: on the grounds that they create noise and dirt, local governments have required that vehicle repair businesses be moved from roads leading to the town to special areas outside the town. This is hard on both the artisans concerned and their clients. Owners must pay to have their vehicles towed out to the mechanics and for taxis in both directions. They also lose security, since a vehicle unguarded overnight may well disappear, whereas if the mechanic lives nearby the area can be secured and watched. Port Harcourt also moved carpenters to a single area where they are isolated with large numbers of competitors rather than available in the neighbourhood. Aba left mechanics and other craftsmen where they were, in the centre of town, providing they keep the area tidy – which they do (Peil, 1986).

There is little impetus to maintain a clean environment in the undeveloped conditions of many large markets – earth floors soon become muddy when there are no roofs to keep off the rain and rubbish accumulates when there is no proper place to put it. Storied concrete markets have been more common in Francophone than in Anglophone cities. They are easier to keep clean, but the attempt to recover the costs of building and maintaining them deprives traders of much of their income. Stall holders often desert the market for the streets, where costs are lower and more customers are to be found.

Street trading is well established in most West African cities but is seen by administrators as a source of 'mess'. In some areas, traders fill the pavements with their goods, forcing pedestrians to walk in the streets. As a result, petty traders are harassed by the police in many cities in an effort to confine them to markets, which are far too small to hold all of them. The popularity of street and house trading is hardly surprising; costs are low enough to provide almost anyone with a profit, and people can buy goods wherever they happen to be. Most traders produce no more street rubbish than the rest of the population, and the provision of containers (as in Aba) would solve the problem without depriving large numbers of women of their livelihood and customers of the goods they need, at a price they can afford to pay.

Conclusion

West African urban environments leave much to be desired, both at the domestic and at the civic level. For the low-income majority, housing is a necessity about which they can do little and services are unreliable at best. Few residents own houses, and for most the town is merely a 'farm' like any other – to be used to produce a living (Aronson, 1978b). They have no particular commitment to the place where they are living and ignore the environment because it is not considered important or amenable to change.

Residential crowding is not seen as a particular problem, since many rural families are nearly as crowded, with even fewer amenities and services. Money is spent on food and clothing rather than on enhancing the family accommodation; any surplus is saved for education or ceremonies or sent home. The environmental preference tends to be a compound swept to the bare soil; shrubs and trees attract mosquitoes. There are cultural and national differences, but generally peoples with a long history of significant urbanisation take much the same approach as others for whom urban life is less familiar. People get on with their lives and leave the environment to look after itself. The link between a poor environment and ill-health is not well understood. They complain of poor water supplies, irregular electricity, unmade roads and muddy markets, but it is like complaining about the weather – no one expects much change. Economic conditions are a far more important cause of popular unrest. In such a situation, government action to improve environmental conditions is not necessarily rewarded at the polls. Given financial constraints, there is little they can do beyond encouraging self-help.

Note

I am grateful to the Nuffield Foundation for grants to carry out several research projects discussed here.

References

Aina, T., 1990, 'Petty landlords and poor tenants in a low-income settlement in metropolitan Lagos, Nigeria', in P. Amis and P. Lloyd (eds), *Housing Africa's Urban Poor*, Manchester, Manchester University Press, pp. 87–101.

Animashaun, I.A., 1981, 'Housing stock: a constraint on urban residential migration in Uyo, Nigeria, *African Urban Studies*, No. 9, pp. 13–29.

Aronson, D.R., 1978a, 'Capitalism and culture in Ibadan urban development', *Urban Anthropology*, 7, pp. 253–69.

—, 1978b, *The City is Our Farm*, Cambridge, Massachusetts, Shenkman Publishing Co.

Barnes, S., 1986, *Patrons and Power; Creating a Political Community in Metropolitan Lagos*, Manchester, Manchester University Press for the International African Institute.

Cline-Cole, R., 1987, 'The socio-ecology of firewood and charcoal on the Freetown peninsula', *Africa*, **57**, pp. 457–97.

—, 1989, 'Inequality and domestic energy in Kano (Nigeria) and Freetown (Sierra

Leone)', in K. Swindell, J.M. Baba and M.J. Mortimore (eds), *Inequality and Development*, London, Macmillan, pp. 243–68.

FOS, 1985a, *National Integrated Survey of Households*, Lagos, Federal Office of Statistics.

—, 1985b, *Social Statistics in Nigeria 1985*, Lagos, Federal Office of Statistics.

—, 1989, *Consumer price index, May 1989*, Lagos, Federal Office of Statistics.

Ghana Government, 1990, *Housing Needs Assessment Study*, Kumasi, Housing and Urban Development Associates, Department of Planning, University of Science and Technology.

Inyang, P.E.B., 1978, 'Environmental pollution in Nigerian towns', in P.O. Sada and J.S. Oguntoyinbo (eds), *Urbanization Processes and Problems in Nigeria*, Ibadan, Ibadan University Press, pp. 169–77.

Konadu-Agyemang, K.O., 1991, 'Reflections on the absence of squatter settlements in West African cities: the case of Kumasi, Ghana', *Urban Studies*, **28**, pp. 139–51.

Lagos State Government, 1989, *Monthly Statistical Indicators*, Part II, Ikeja, Statistics Division.

Master Plan for Metropolitan Lagos, 1978, Bulletin 3, Ikeja, Lagos State Ministry of Works and Planning.

—, 1980, vol. I and II, Lagos, Wilbur Smith and Associates for the State Ministry of Physical Planning and Land Matters.

Ogionwo, W., 1979, *A Social Survey of Port Harcourt*, Ibadan, Heinemann Educational Books.

Pasteur, D., 1979, 'The Ibadan comfort stations programme: a case-study of the community development approach to environmental health improvement', *Journal of Administration Overseas*, **18**, pp. 46–58.

Peil, M., 1976, 'African squatter settlements', *Urban Studies*, **13**, pp. 155–66.

—, 1981, *Cities and Suburbs: Urban Life in West Africa*, New York, Holmes and Meier.

—, 1986, 'Sanitation versus small business', *West Africa*, 3580, p. 777.

—, 1991, *Lagos: The City Is the People*, London, Belhaven Press.

Stren, R., 1985, *Two Nigerian Towns in the Eighties: A Socio-Economic Survey of Idah and Makurdi, Benue State*, Project Ecoville, Working Paper No. 21, Toronto, Institute for Environmental Studies, University of Toronto.

Sule, R.A.O., 1978, 'An assessment of the Nigerian housing allocation policy: the case for the ballot system', *African Urban Studies*, No. 1, pp. 67–85.

—, 1981, 'Environmental pollution in an urban centre: waste disposal in Calabar, *Third World Planning Review*, **3**, pp. 419–31.

Tipple, A.G., 1982, 'Housing in Kumasi; a survey of households in low income areas', Research Report 4, Kumasi, Department of Planning, University of Science and Technology.

Tipple, A.G. and Hellen, J.A., 1986, 'Priorities for public utilities and housing improvements in Kumasi, Ghana'. Seminar, Paper 44, Department of Geography, University of Newcastle upon Tyne.

West Africa, weekly, London/Lagos.

11 Housing and environment in an Indian city: a case of a squatter settlement in Allahabad, India

Harikesh N. Misra

Whatever the scenario of development, urbanisation is generally considered as an important indicator of socio-economic change. The implications and consequences of large-scale urbanisation are associated with a familiar catalogue of social and economic problems – spectacular growth of shanty towns; inadequate water supply and sewerage infrastructure; high rates of unemployment and proliferation of casual workers making out a marginal existence in petty production and trading; crime, traffic congestion' (Gore, 1984, p. 55). In the present context housing and environment issues pose major urban problems for India. Planners, policy-makers, administrators and academics have long been concerned with the problems of inadequate shelter and deteriorating quality of life for the majority of the population living in the cities. But the magnitude of the problems has become much more complex. The rapid growth of population and increasing rate of unskilled and semiskilled migrants from rural to urban areas have been adding to the problems of over-crowding and unemployment in towns and cities. The consequences of too many people sharing too few resources are obvious: 'One cannot fail to view the growth of squatter colonies, the congestion, the shortage of solid shelter and near break-down of municipal services and transport systems in any metropolitan region' (Koenigsberger, 1976, p. 14; see also Misra, 1992).

Migrants, it is estimated, constitute about 50 per cent of the population of the 'million' cities of India. In cities having a population of 300,000 or less the share of migrants is between 30 to 40 per cent of their total population (Misra, 1979, p. 32). 'A vast majority of this population is denied the minimum basic needs of life such as shelter and access to water and other basic needs like food and clothing' (Sethuraman, 1978, p. 2). More than 30 per cent of the country's urban population lives in '*Jhuggies*', and it is likely they will outnumber those living in properly built houses by the year 2000.

The purpose of this chapter is to examine the state of housing and environment in Indian cities by analysing the growth and development of Allahabad and one of its squatter settlements, Chheetpur.

National scene: an overview

Indian cities are facing acute and unprecedented problems of housing and environment as the country passes through a period of rapid urbanisation. Despite being only 25 per cent urbanised India's urban population is the fourth largest in the world after China, Russia and the United States. According to the 1991 census the total urban population of the country is 217.18 million distributed in 3,696 urban centres which are subdivided into six classes. A characteristic feature of India's urbanisation is over-concentration. This is demonstrated by the fact that during a period of ninety years (1901–91) the urban population increased nine times from 25.6 million to 217.18 million, whereas the number of urban centres has only doubled from 1,834 in 1901 to 3,693 in 1991.

The urban population is expected to reach 326 million by the year 2001 adding further to the process of overcrowding. The uneven distribution is again exemplified by the predominance of 300 Class I urban units (referred to as cities) which encompass about 65 per cent of the urban population. It is this class which has recorded the highest growth (46.8 per cent in 1981–91). The problems of overcrowding and concentration are further accentuated because 50.57 per cent of the population of Class I urban centres are located in twenty-three metropolitan centres and more than half of this (26.64 per cent) is found in four megacities – Bombay, Calcutta, Delhi and Madras. This pattern of distribution of urban population has several environmental impacts which can be observed most clearly in the housing problems facing these cities.

Housing not only satisfies physiological and psychological needs but also provides protection from diseases and accidents. In short, healthy housing conditions contribute to a healthy urban environment. Unfortunately, a large proportion of the population, especially low-income groups living in slums and squatter settlements, have been deprived of access to healthy housing conditions and a satisfactory environment. While it is true that over the last four decades the shortage in dwelling units has been increasing at a rapid rate both in rural and urban areas, the problem of rural housing has never been as serious as that of urban housing. Several programmes for housing the urban poor have been launched from time to time, but the housing situation has deteriorated, particularly in the cities where large-scale migration from rural to urban areas has been most apparent. Consequently, *Jhuggies, Jhopadis, Jhopadpattis, NaiBastees* or shacks made of tin, wood, thatch, rags, mud and canvas continue to be characteristic features of the Indian urban landscape. The National Building Organisation in 1981 estimated the shortage of housing units in urban areas to be seven million; this is expected to reach 15.5 million by the year 2000 (National Research Development Corporation, 1992, p. 367).

While the shortage of dwelling units is depressing in itself, the quality of existing urban housing is also a major cause for concern. Urban housing conditions are summarised in Table 11.1. This table clearly reveals the cramped conditions of the dwelling units and the poor infrastructural facilities during 1961–81. These figures present an alarming picture. More than half of the urban households lived in one-room units. Water, electricity and latrine facilities, which are basic amenities, were not available to nearly one-third of the total urban households. In particular the conditions of housing and environment in the four megacities are of an appalling quality. Thus 'the majority of Bombay's population lives in one room houses and a shortage of 55,000 units is added yearly to an 800,000 backlog and if the present trends continue, by the year 2000, 75 per cent of Bombay's population will be living in slums' (CSE, India, 1989, p. 13). In Calcutta, for example, 77 per cent of all families share lavatories with other families and

Table 11.1 Urban housing conditions in India

Description	1961	Year 1971	1981
1. Average number of persons per room	2.62	2.77	
2. One-room dwelling			
a Per cent of households	53.05	50.10	
b Per cent of population	43.52	41.72	
3. Two-room dwelling			
a Per cent of households	24.66	26.93	
b Per cent of population	26.14	28.08	
4. Infrastructure per cent of household with[1]			
a Tap water[2]	50.67	68.89	79.66
b Electricity			53.48
c Latrine	56.37	60.49	66.81
5. Rented accommodation per cent of household	53.8	52.9	

Notes: [1] 1961, 1971 and 1981 refer to the years 1953–4, 1964–5 and 1973–4, respectively.
[2] Refers to water available through taps, tubewells and hand pumps

Source: India (1980)

more than 10 per cent have no facility at all (Gilbert and Gugler, 1982, p. 26). The housing and environmental problems of Delhi, Madras and Kanpur are similar and other industrial cities are no better served.

In the following section, the city of Allahabad is examined in some detail in order to understand, at a local scale, the factors underpinning the processes responsible for the present state of housing and environment within the broader Indian context.

The case of Allahabad

Pre-Muslim in origin, the city of Allahabad lies at the confluence of the Ganga, the Yamuna and the invisible Saraswati in the mid-Ganga plain of the state of Uttar Pradesh. It is a central place and a cultural centre, exhibiting both sacred and secular geographies. The sacred geography is periodically manifested when it becomes the site of the largest pilgrimage in the world every twelfth year.[1] Because of its strategic location it was made capital of the United Provinces during British Rule which boosted its administrative importance. With the opening of Allahabad University in 1887, the fourth in the country after Bombay, Calcutta and Madras, it also became an important educational centre. Although it suffered a setback when the capital of the province was shifted to Lucknow in the 1920s, it still stands second only after Lucknow because most of the offices including the High Court continue to be located here. A few industrial concerns such as Indian Telephone Industry, Swadeshi Cotton Mill, Triveni Engineering, Glass Factory, Geep Torch Factory, for example, have also located here during the last three decades.

Socio-economic profile

Allahabad city is one of the five major cities in the state of Uttar Pradesh. The other four cities are Kanpur, Lucknow, Agra and Varanasi. All these have recorded similar growth rates except Kanpur which is the major industrial city of the state. During the last decade (1981–91) the growth of population in Allahabad was only less than Lucknow, the state capital. The population of the city of Allahabad has increased by about five times during 1981–1991 and is likely to reach 1.26 million by the year 2000. It became a Municipal Corporation in 1960, when Bamrauli, Phaphamau and Naini were added to the city map of Allahabad (Figure 11.1). The area of the city increased from 70.05 square kilometres to 82.18 square kilometres in 1971, but the jurisdiction of Allahabad Development Authority, which came into existence in 1974 to execute the master plan, extends beyond the Corporation limit to a radius of eight kilometres.

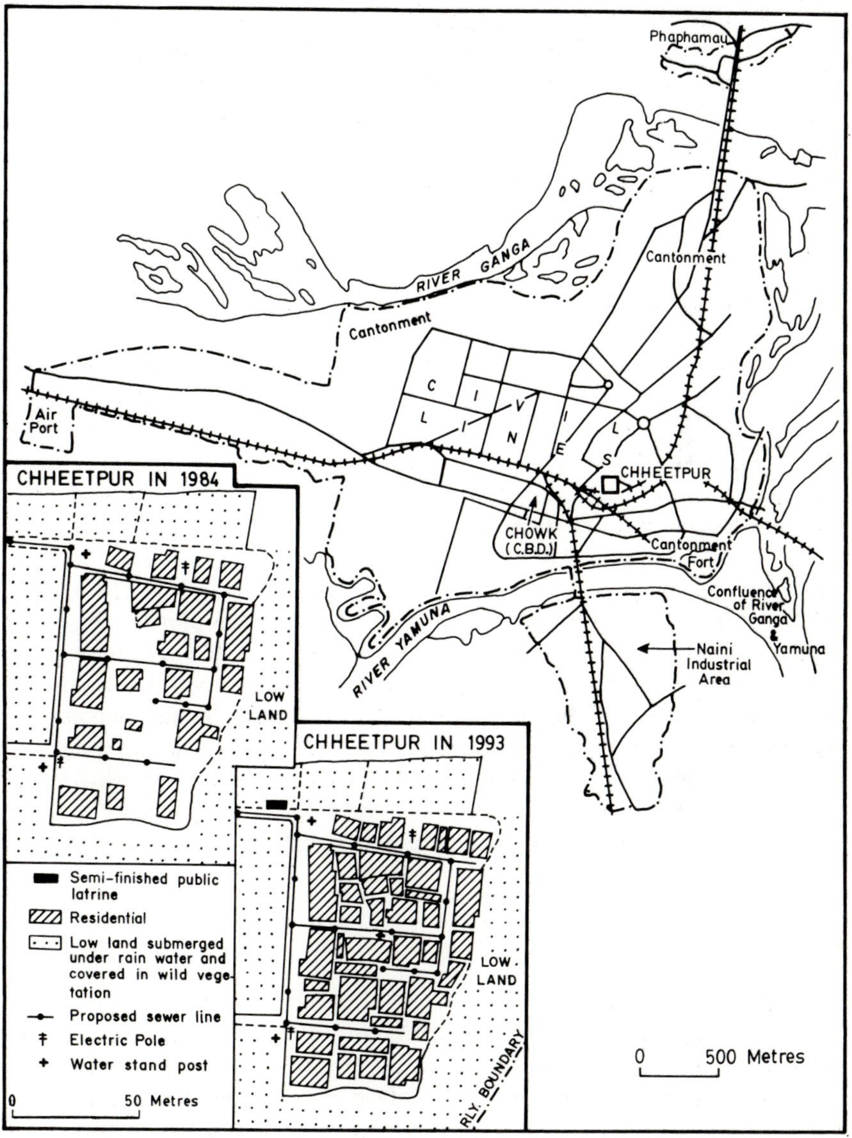

Figure 11.1 Allahabad City and Chheetpur squatter settlement

Allahabad is a city of migrants, and in-migration plays a very significant role in spurring the growth of population. This is further substantiated by the low sex-ratio which is 816 females per 1,000 males according to the 1991 census; the sex-ratio has been highly imbalanced since 1941. The migrants are the agricultural labourers who continue to have their roots in

Table 11.2 Occupational structure of Allahabad City (%)

Sectors	1921	1951	1961	1971	1981
Primary	9.4	4.0	3.7	5.7	4.1
Secondary	14.6	23.3	24.6	22.2	28.2
Tertiary	76.0	72.7	71.7	72.1	67.7

Source: Census of Allahabad District, 1921–81

Table 11.3 Distribution of workforce in tertiary sector of Allahabad (%)

Particulars	1921	1951	1961	1971	1981
Commerce & trade	14.8	18.3	16.6	19.6	13.2
Transport & storage	7.3	10.9	12.8	9.3	13.4
Other services	53.9	43.5	42.4	43.2	41.1

Source: Census of Allahabad District, 1921–81

villages (Misra, 1981). In addition, the economic base of the city is very weak. Production-orientated functions are few and are relatively recent phenomena, appearing only during the last twenty to thirty years. The tertiary sector provides most employment opportunities (68 per cent). In this sector 'other services' are most important (Tables 11.2 and 11.3). This category includes people engaged in the informal sector which accounts for more than 41 per cent of the total work force. Surprisingly, employment in the secondary sector has increased significantly from 22.2 per cent in 1971 to 28.2 per cent in 1981.

Housing and environment

The conditions of housing and environment, both in terms of quality and quantity, are in a deplorable state in the city of Allahabad. About 80 per cent of the population live in substandard dwellings; 44 per cent live in one-room apartments, 31 per cent in two rooms, 12 per cent in three-room flats, six per cent in four-room houses and two per cent in five-room houses. It is only a small proportion (three per cent) which has the luxury of living in houses with five rooms or more.

The housing shortage increased to 90,000 in 1984 compared with 60,000 in 1977. Clearly, the shortage of housing may increase further if construction rates do not accelerate. It is estimated that by the year 2000, the population of the city and its environs may increase to 1,267,227 and housing requirements may exceed 492,000 units. While the city population growth rate was 93 per cent during 1951–81, the construction rate for housing was only 26 per cent during the same period. The construction work undertaken by the Allahabad Development Authority and other organisations presents a dismal picture in view of the demand for shelter. This deficit apart, infrastructural facilities such as drainage, sewage, potable water, sanitation and other public utilities are very poorly distributed. Nearly 62 per cent of houses do not have a separate kitchen; potable water supply is available only to 60 per cent of the households and a flush latrine is a luxury of only a privileged few (34 per cent of the households). Only a quarter of the city is covered by sewerlines. Water, air and noise pollution have been causing severe health problems. The water of the river Ganga is becoming polluted owing to discharge of sewage and sullage effluents and also from toxic waste materials from the Indian Farmers' Fertilizers Corporation located 25 kilometres upstream of Allahabad in Phulpur. The cremation of dead bodies at Sangam and Rasulabad has also been causing pollution.

A city which was once famous for its clean environment has undergone a dramatic change during the last two decades. Roads choked with traffic are common features and even the piped water supply gets contaminated because of leakage, promoting the spread of several water-borne diseases. Air and noise pollution are increasing owing to the multiplicity of three-wheelers which emit fumes, smoke and lead on almost all major roads within the city. Even the civil lines (the residential area reserved for Europeans in the colonial city) which have been areas of relatively high environmental quality since the days of the British Empire are not unaffected by this kind of environmental pollution.

Slums and squatter settlements

The infrastructure of the city is quite simply unable to cope with the increasing population. Lack of access to land and infrastructural facilities by poor income groups poses serious problems of housing and environment quality. It is infested with slums and squatter settlements both in the central and peripheral areas. The population of these settlements ranges between 500 to 5,000 people. In 1981–2 the population of slums and squatter settlements was estimated to be about 188,000, which has now risen to 300,000. These settlements are typical examples of popular settlements (Misra, 1988): they are illegally built; there are no building norms; tenure rights are questionable; services and infrastructural facilities are inadequate and sites are breeding grounds for several epidemic diseases; and they are water-logged for much of the year. Other locations for these settlements are vacant lands near railway stations, bus stations and market-centres, for example, and the dwellings are built of mud, wood, tin, rags, thatch and canvas. The income level of the residents is low and the size of the household is between five and ten persons or even more. Although, over time, services by the municipal government under the slum improvement programme are provided, these are usually paltry and inadequate.

Several factors contribute to this crisis of illegal development. First, the geographic location of the city between the two rivers means that physical expansion is difficult. Second, the city is situated in the socio-economically backward region of the northern plain of India where agriculture is the principal occupation and population density is very high (600 persons per square kilometre). Consequently, large-scale migration of rural people adds significantly to the crowding and congestion of the city. Third, the implementation of the 1976 Urban Land Ceiling and Regulation Act, which was enacted to force urban land on to the market has been ineffective. This provides a ceiling of 1,500 square metres on the holding of vacant land in urban areas. A number of factors have contributed to the failure of the Act to work in practice; however, the main reason has been the large-scale intervention by the judicial courts. Fourth, the Allahabad Development Authority, which is responsible for providing housing to low-income group communities, has failed in its efforts to counteract speculation in the urban land market. Private landowners are selling land to cooperative housing societies. These housing societies, about 200 in number, are doing good business by selling and purchasing the land, adding to price speculation. As Kumar notes: 'Since land is the dominant component of cost of the housing today access to land is the central problem in the provision of shelter' (1989, p. 1285). Consequently, the low-income group derives little benefit from these societies. In addition there is an inadequate number of programmes for the construction of new houses which is compounded by inadequate loan facilities. Furthermore, the poor economic conditions and the increasing number of immigrants have further deepened the crisis. Fifth, the land-tenure system of the city is complex, with much of the land still held under a system imposed by the British. The city's land use may broadly be divided into agricultural and non-agricultural categories. The tenure system of agricultural land is less complicated in comparison to non-agricultural land (Figure 11.2). There are four principle types of non-agricultural land tenure: freehold, *Nazul*[2] government estate[3] and lease (including several sub-divisions). Most of the land comes in the category of Nazul (8.5 per cent) or government estate (39.4 per cent). The cantonments occupy 22.3 per cent which was meant for defence purposes. The freehold land, which is easily transferable, is just 28.6 per cent of the total municipal land and is easily beyond the reach of low-income groups. The powers of transfer of *Nazul* or government land are vested with the state government, but in the absence of a clearly defined land policy, the situation is static; even leases are not being renewed. Nevertheless, the problem could

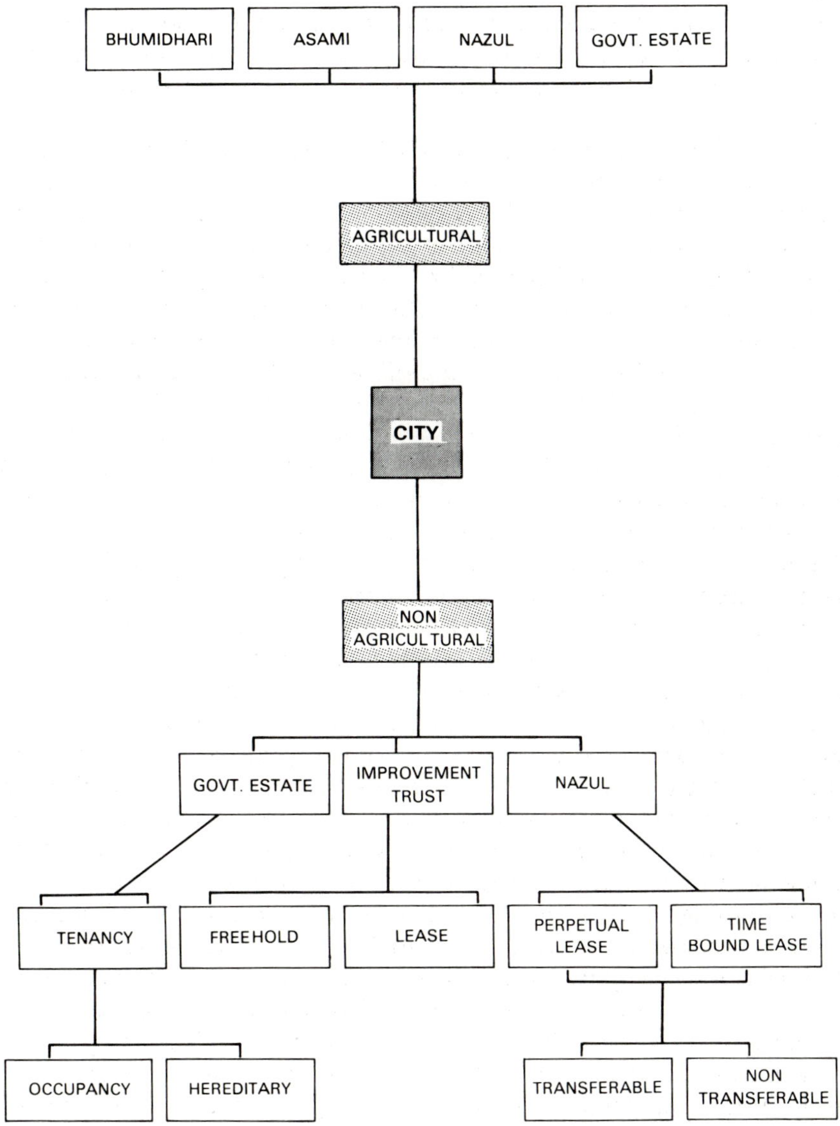

Figure 11.2 Land tenure system of Allahabad City

be partially eased by releasing some land on to the open market. Sixth, building codes and norms have been fixed by the planners and architects and the reality of the poor building for themselves has not been given due consideration. The provision of cheap and affordable houses is, therefore, made difficult by rigid adherence to municipal by-laws.

In the following section, a case study of Chheetpur squatter settlement is presented in some detail so as to chart the changing conditions of housing and environment in a self-built community.

Chheetpur: a squatter settlement

Otherwise called shanties, *Jhuggies* and *Jhopadies*, squatter settlements have now become an integral part of any Indian city. They are indicative of the gap between basic human needs and institutional capabilities to meet these needs. When society fails to respond to the housing needs of the people they find ways, often illegal, to settle themselves. Squatter settlements are products of these spontaneous responses to institutional rigidities and, one may even say, callousness. They spring up on vacant land earmarked for public purposes, private land left unbuilt for speculation, low-lying marshy lands or waterfronts. Squatter settlements are also frequently designated as slums. For the purposes of this chapter, slums are defined as the overcongested old quarters of the city interior which are dilapidated, derelict and inhabited by low-income families. Squatter settlements, in contrast, are newly settled areas largely inhabited by migrants who can neither afford to buy land nor build houses following the municipal by-laws. In both the cases the end results in terms of human misery are the same, but the human ecology of the two are different.

The squatter settlement of Chheetpur came into existence some fifty years ago. It is a dynamic settlement, although the changes which have taken place have not been dramatic (Misra, 1988, reports the findings of a survey conducted in 1984). A reconnaissance survey was conducted at the beginning of 1993 to chart the changes which have taken place since 1984 in housing, environment and quality of life (see Figure 11.1). Occupying an area of less than a hectare along the railway line to the southern part of the city, it now contains several new houses with improved structures. Although the surroundings of the site continue to be flooded during the rainy season, the dwellings of the residents are no longer inundated.

Historical background

According to the elderly inhabitants of the Chheetpur, their forefathers came from rural areas to work as unskilled construction workers, building mud houses on the government land close to the present day Swarup Rani Hospital. However, when the 'civil lines' area of the city was developed, they were evicted, and the only indication of their former settlement is the remnant of a masonry well built for drinking water. A proportion of the evicted families moved to the vacant land of the Kayasth Patshala Trust.

They tilled this land on behalf of the Trust, with the women working as domestic servants and vegetable sellers, while the children looked after goats, buffaloes and pigs.

Forty years ago the Kayasth Patshala Trust decided to improve the campus of their college and the residents were once again forcibly evicted. Five families moved to the present site which was swampy and low lying. Indeed, the site was so inhospitable that the authorities were unconcerned about the new arrivals. Thatched huts were erected using local materials and the Trust levied Rs.1 per household as rent for the land. During the Second World War the Indian Hume pipe factory was set up close to this settlement, and a number of the settlers were employed there. The size of the settlement was increased by post-war rural–urban migration.

The Kayasth Patshala Trust which owns the land allegedly used its students to set fire to new huts. However, despite this intimidation new huts have been constructed. The settlers responded to this harassment by vertical expansion, building two- or even three-storey structures. Furthermore they have stopped payment to the Kayasth Patshala Trust, claiming title to the land based on long-standing tenancy. However, this matter has not been settled and has been in the courts since 1974.

A voluntary organisation called the Harijian and Backward Committee was constituted in 1980 and yet another organisation called Chheetpur Vikas Samiti (Chheetpur Development Committee) was formed in 1983. These organisations succeeded in arousing a political awakening in the people, and the facilities they have been able to obtain have been a result of the efforts of these organisations. However, these organisations no longer exist, their original functions being superseded by other needs.

People and their livelihood

At the time of the first survey in 1984 the population of Chheetpur was 492: 281 males and 211 females. The population structure was marked by a high dependency ratio, 42 per cent of the people were children aged fourteen and under and 6 per cent were aged over sixty (Table 11.4). Only very marginal changes in the age and sex structure have occurred in the intervening years. The population growth is as high as 3.5 per cent per annum, brought about by a combination of a high birth-rate and declining death-rate and the impact of new migrants.

The social structure of Chheetpur has lost its original rigidity but continues to be loosely tied to the past. The majority of the inhabitants belong to what is now termed 'the backward castes'; scheduled caste[4] households are next. Higher castes are in the minority as are the Muslims. The percentage of upper-caste households increased in 1993 (Table 11.5). These are tenants who have opted to live in the settlement in the absence of any

Table 11.4 Age and sex structure of population in Chheetpur, Allahabad, 1984–93 (%)

Age group	1984 Male	1984 Female	1984 Total	1993 Male	1993 Female	1993 Total
0–15	43.4	40.3	42.1	40.2	36.6	38.6
16–59	51.6	51.7	51.9	55.3	57.0	56.0
60+	5.0	8.0	6.0	4.5	6.4	5.4
Total	100.0	100.0	100.0	100.0	100.0	100.0

Source: Field Work, 1984–93

Table 11.5 Households in Chheetpur according to caste, 1984–93 (%)

Caste	1984	1993
A. Upper caste	2.8	5.7
B. Backward caste	67.6	68.2
C. Scheduled caste	25.9	25.0
D. Muslim	3.7	1.1
Total	100.0	100.0

Source: Field Work, 1984–93

other alternatives. The improved housing conditions have also started to attract some student tenants. The rents from tenants are becoming a reasonable source of income in the case of at least ten households in the settlement. Although it is possible to jump the caste barriers to earn a livelihood, caste and occupation linkage is still a significant feature. Only one-third of the working population is employed in the formal sector; the remainder is engaged among the informal activities of the city. Activities such as vegetable vending, tailoring, masonry, carpentry, pottery and rickshaw-pulling constitute the main sources of income. In comparison with the situation in 1984, the employment structure has changed little – the most notable feature is that the number of people employed in the informal sector has increased.

High dependence on low-wage employment keeps the per capita income of Chheetpur residents at a low level. The upper-caste residents and Muslims feel that they are better off than their counterparts in the villages they have left, but surprisingly they live only in rented houses. Apparently, they feel little attachment to the settlement and hope to leave it as and when they have enough money to buy or build a house elsewhere. This is also clear from their limited involvement in community activities.

Living environment

The living environment of Chheetpur can be bluntly described as depressing and inhuman. The low level of sanitation spreads diseases, which find ready victims in under- and malnourished Chheetpurians. There is a degree of social cohesion and fellow feeling because of the interdependence of the lives of residents, but fractional and family quarrels are equally apparent.

Housing The housing conditions are bad in all respects. One- or two-room houses shelter five to seven members – including adult males, females and children, with each room measuring not more than nine to twelve square metres. Building materials in Chheetpur are locally derived, with mud being used for walls and tiles, and bamboo and straw for thatch. In 1984 65 per cent of the houses were mud structures and 26 per cent were brick-walled with tiled roofs. Only 9 per cent of the structures were made of baked bricks and cement. This situation, however, has changed significantly (Table 11.6). Mud-walled houses now represent only 19 per cent of the total stock, while cemented houses and brick-walled houses with tiled roofs account for 40 per cent and 41 per cent respectively. This reveals a remarkable capacity for self-builders to improve their housing conditions. Almost every house begins with one room; as money is accumulated and the size of the family increases, incremental expansion takes place. This is one reason why nothing can be planned effectively in advance. None of the houses has a separate kitchen and lavatory, and defecation occurs outside along the railway track or other open spaces or in open drainage lines on the two sides of the street. A public latrine was being constructed at the initiative of the Municipal Corporation, but it came to a halt because the residents objected to the selection of the site. In addition a sewer line was proposed nearly a decade ago, but this has not yet become a reality. Animals and human beings live together, causing

Table 11.6 House types in Chheetpur, Allahabad, 1984–93 (%)

House Type	1984	1993
1. Cemented	9.3	39.7
2. Brick-walled with tiled roofs	25.9	41.4
3. Mud-walled with tiled roofs	64.8	18.9
Total	100.0	100.0

Source: Field Work, 1984–93

major problems of sanitation. The inhabitants have become so used to this environment that individual and collective efforts to improve the quality of life are virtually non-existent. Everyone is resigned to the present situation, and any improvements depend on the intervention of outside agencies.

Chheetpur did not have a protected water supply until 1970. Prior to this it depended on a well whose water was polluted because of the water-logging around the settlement. The first municipal water supply came in 1972 when one public tap was installed, and this was followed by another four years later. These two taps were the only source of water for all the residents; a third water tap was fitted only in 1990. Although electricity came in 1978 there are only two street poles, and almost every house has illegal connections.

Health Chheetpur is situated very close to the largest government hospital in Allahabad, but its inhabitants rarely use its services. Hospital services are free, but the residents do not receive attention without a long wait and without paying unofficial fees. Other options are available. Thus the more wealthy among the Chheetpurians consult private doctors, while the majority depend on a homeopathic doctor deputed by Satya Sai Mission, a voluntary agency. In addition a few families consult with witch-doctors and faith healers known as '*Ojha*', in the absence of early detection and treatment of diseases, the health of the sick deteriorates rapidly. Nevertheless, given the environment, housing, medical services and other facilities such as water supply, the health conditions of the people could be worse. Although one can see a number of healthy people, those suffering from ill-health constitute the majority of the population.

The most common diseases of Chheetpur are jaundice, diarrhoea, dysentery, typhoid and cholera. Infestations of amoeba, round worm, gunea-worm, hookworm and thread worm are also common. Clinical tests show that 50 per cent of the people suffer from worm infection. Cases of anaemia, joint paints, dyspepsia and eosinophilia have also been observed, while tuberculosis and burn injuries are not uncommon. Burns are usually a consequence of overcrowding and poor housing conditions. Poisoning results from contaminated food and illicit liquor, tuberculosis from un-hygienic conditions and poor diet, and cancer from smoking and liquor consumption.

Results of the diet survey converted into calories are given in Table 11.7. This table shows the caloric intake of males and females separately, high-lighting not only the poor diet but also the differences that exist between male and female diets. The disadvantaged position of females reflects the broader societal value system in which males are preferred to females. In summary, these data demonstrate the close link which exists between the health of the people and the environmental setting of Chheetpur. The lack of medical facilities is certainly one factor which keeps the people in poor

Table 11.7 Caloric intake of male and female population in Chheetpur squatter settlement, Allahabad (%)

Caloric value	Male	Female	Total
Below 500	6.0	7.6	6.7
500–749	17.4	13.3	15.6
750–999	13.5	20.8	16.6
1,000–1,249	22.4	39.8	29.8
1,250–1,499	20.0	12.8	16.8
1,500–1,749	14.2	4.7	10.1
1,750–2,000	5.0	1.0	3.6
Above 2,000	1.5	–	0.8
Total	100.0	100.0	100.0

Source: Field Work, 1993

health. However the lack of such facilities refers only to the curative dimension of health care. Much more significant is the preventative dimension which suggests that so long as poor diets and a polluted living environment persist, so will chronic ill-health.

Education The people of Chheetpur are adjacent to several colleges and schools, but only 30 per cent were literate in 1984; however by 1993 the literacy rate had increased to 37 per cent. This is mainly because of a very significant increase of literacy in the upper-caste community. In contrast the backward and scheduled caste communities show only very marginal improvements in literacy. Preoccupation with meeting everyday needs means that investment in the future receives a low priority. The school which the children attend has practically no facilities; parents have no interest and many children go only at the time of admission. This is compounded by the fact that the environment at home is opposed to formal education, an attitude that is reproduced through the generations. It is not the lack of educational facilities alone that keeps the literacy level of Chheetpur so low. Some people do send their children to school. But why not the others? These others are the poor, the sick, the unmotivated and the men and women who do not aspire to anything beyond what they have. In fact they lead their life from day to day; for them today is the future.

Conclusion

Without doubt the Chheetpurians have a tremendous capacity for change, innovation, adjustment and compromise. But they are also inward-looking, aware of the outside world but not knowing how to enter it. The city which attracts the rural poor provides a limited and uncertain future for Chheetpurians. The reasons for this malaise are complex and multiple in character. On the one hand the large number of in-migrants has led to overcrowding and to major problems of urban development and planning. On the other hand, however, the combination of institutional rigidity and limited capabilities have hampered the development of effective and imaginative planning. This is apparent in the context of both the urban fabric in general and in the failure to create and implement effective strategies for tackling the crucial relationships between housing and environment quality.

Notes

1. Every twelfth year more than five million pilgrims congregate here. Each year during the Hindu month of *Magh* (January), more than one million people travel from all parts of India to bathe in the sacred confluence.
2. *Nazul* land is the land which was originally unclaimed and then taken by the colonial government and given to people for cultivation without written title.
3. Government estate land is the land which was confiscated by the British from the landowners who fought against them in 1857.
4. The 'scheduled' castes, together with the scheduled and other tribes constitute the lower third in economic terms and are determined for each state.

References

Centre for Science and Environment (CSE), 1989, 'The environmental problems associated with India's major cities', *Environment and Urbanisation* **1**(1), pp. 7–15.

Gilbert, A. and Gugler, J., 1982, *Cities, Poverty and Development: Urbanisation in the Third World*, London, Oxford University Press.

Gore, C., 1984, *Regions in Question, Space, Development Theory and Regional Policy*, London, Methuen.

India, Government of, 1980, *India, A Reference Manual*, Delhi, Government Printers.

Koenigsberger, O.H., 1976, *The Absorption of Newcomers in the Cities of the Third World*, Mysore, Institute of Development Studies.

Kumar, A., 1989, 'National housing policy: the implications', *Economic and Political Weekly*, **23**, pp. 1285–94.

Misra, K., 1992, *Housing the Poor in Third World Cities: Choice, Behaviour and Public Policy*, New Delhi, Concept.

Misra, H.N., 1981, 'Rural roots of urban poor: a case study of informal sector in an Indian city', in R.P. Misra (ed.), *Rural Development: National Policies and Experiences*, Singapore, Maruzen Asia, pp. 215–29.

—, 1988, 'Popular settlements in the city of Allahabad', *Cities*, **5**(2), pp. 162–83.

Misra, R.P., 1979, *Habitat Asia*, Volume 1, New Delhi, Concept.

National Research Development Corporation, 1992, *Invention Intelligence*, New Delhi, December.

Sethuraman, S.V., 1978, 'The informal urban sector in developing countries', in A. Desouza (ed.), *The Indian City*, Bombay, Manohar.

12 Urban environmental controls and low-income housing in southern Africa

Deborah Potts

The most significant urban problems in Africa tend to differ from the classic problems prevalent in more urbanised regions of the world. Although problems associated with industrialisation and traffic exist, the most pressing human-produced environmental issues are associated with the provision of shelter. This is still true for southern Africa, despite relatively high levels of urbanisation in several countries and the development of a number of fairly large cities, particularly in South Africa.

A significant factor in urbanisation in southern Africa has been the very high level of planning and control associated with urban development in some countries. This is true of all the cities discussed in this chapter: Harare in Zimbabwe, Lilongwe in Malawi, Gaborone in Botswana and Durban and Johannesburg in South Africa. However, despite the often very heavy-handed attempts to control the urban environment, rapid population growth has not been matched by sufficient planned housing provision to prevent the emergence of environmental problems. These include overcrowding within both the pre-existing planned housing stock and the newer unplanned or irregular settlements and problems associated with a lack of proper sanitation. This chapter will focus in particular on those problems in the selected cities and the planning response to them.

Overcrowding and sanitation problems in low-income settlements

Descriptions of environmental problems in low-income settlements often concentrate on the irregular nature of the materials used in the buildings: for example packing cases, corrugated iron or traditional mud bricks. However, this focus is often misplaced since these may not be a major source of discomfort or danger for the inhabitants. Rather, when considering the relationship between health and the environment of human settlements, the most striking features of poor housing are overcrowding and lack of basic sanitation (Novick, 1976, p. 30). In unplanned settlements with a high density of dwelling units and people, the problem of the safe disposal of human body wastes is a major health issue. The low population densities found in most African rural areas may make simple pit latrines or even an absence of specific sanitation structures reasonable responses to the problem – but in urban areas different techniques are needed, given that 'human excreta is one of the most dangerous substances with which people can come into contact' (ibid. p. 31). Lack of proper sanitation can help to cause and spread diseases which are among the leading causes of death and disability in the developing world, including the enteric diseases like cholera, typhoid, dysentery, the diarrhoeal diseases and virus diseases such as hepatitis. Intestinal parasites which adversely affect nutrition are also closely associated with poor sanitation. Research in developing countries has shown that morbidity and mortality from such diseases are much higher in low-income settlements without adequate sanitation than in higher-income settlements with better facilities (Bradley, *et al.*, 1992; Pryer and Crook, 1988; Sinnatamby, 1990). Thus sanitation for high-density low-income settlements has been identified as 'the single most important challenge to those concerned with environmental health in the Third World in the 1990s' (Sinnatamby, op. cit., p. 132).

In most low-income urban settlements in Africa there is no piped sewerage system, and indeed some entire cities have no water-borne piped system (for example Kinshasa in Zaïre (Hardoy and Satterthwaite, 1992)). Many urban people have no access to any sanitation system at all, although the data suggest that African cities are somewhat better off in this regard than many Asian cities. In 1985 27 per cent of African urban dwellers had no sanitation coverage, compared to 67 per cent in South-East Asia (Sinnatamby, op. cit.) and 33 per cent in India (Hardoy and Satterthwaite, 1992). The lack of adequate sanitation is due to a variety of factors. The poverty and relative powerlessness of urban governments in Africa form part of the explanation at the institutional level (Stren and White, 1989). At the community level poverty may prevent people from affording even the simplest low-cost pit latrines, but lack of security of tenure, difficult terrain (for example steep slopes or liability to flooding) and sometimes ignorance of the techniques also play a part.

Overcrowding within dwelling units is also a symptom of poverty, since land and extra rooms both have a cost. The levels of overcrowding in different African cities vary. While nearly all major towns suffer from a housing shortage, and some overcrowding will always occur as poor families economise on rents or building costs, the most serious problems occur where the gap between the cost of living space and the price the poor can pay is highest. In some cases overcrowding is caused by the city's location which restricts its areal expansion, but more generally in Africa the availability of land for the urban poor is influenced by the nature of the ownership of peri-urban land and government attitudes towards unplanned settlements. At one end of the spectrum are small urban areas surrounded by land under traditional communal tenure where access to land or rooms can be negotiated with traditional leaders at an affordable cost and where local government either ignores breaches of planning regulations or does not have the capacity to pursue them. At the other end of the spectrum are large cities where land tenure is mainly private, and there is strong government commitment to upholding the rights of private landowners and maintaining a planned and controlled urban environment. This latter scenario will not necessarily prevent illegal and unplanned settlements emerging, but it will tend to constrain their development and consolidation. It is this scenario which has characterised urban developments in South Africa, Zimbabwe and Namibia and the new capital development of Lilongwe, although as will be seen, changing political circumstances and the sheer pressure of population growth have inevitably led to these controls breaking down to some degree. Nevertheless illegal and unplanned settlements are still viewed unfavourably and are rarely secure, exacerbating the overcrowding problem.

Most discussions of overcrowding focus on physical health issues. Small, poorly ventilated dwellings promote the spread of communicable diseases, and overcrowding within them further increases the frequency of transmission of diseases (WHO, 1988). Diseases associated with overcrowding include upper-respiratory tract diseases, measles, TB, meningitis, influenza, mumps, diphtheria and whooping cough. Household accidents are also more common, and lack of space makes it more difficult to maintain hygiene standards generally, including in food preparation (ibid.; Hardoy and Satterthwaite, 1992). The measurement of overcrowding is usually expressed as an index of the average number of people per habitable room. Hardoy and Satterthwaite (ibid.) report that in low-income residential areas in Third World cities there is often an average of four or more persons per room. Recent data for African cities are scarce but a few examples can be given. Room densities in Takoradi, Kumasi and Accra in Ghana in the mid-1970s averaged around 2.5–3.2 (Gilbert, 1992); in Port Harcourt in Nigeria in the same period they were 2.4 (Izeogu, 1989); in Dar es Salaam in 1987 a survey found an average of 2.4 (Kulaba, 1989); in Lagos in 1981 there were 4.1 people per room (Ayeni, 1981); and in the rather special circumstances of Khartoum in 1990, with a huge displaced refugee population, room occupancy rates in informal settlements of fifteen people have been reported (Omer, 1990).

It is important to recognise that Western expectations with regard to sanitation and space requirements often have little meaning in the context of African urban low-income settlements. As Peil (1976) has pointed out, room occupancy rates are not so significant where much time is spent outside in tropical countries. Furthermore 'to a hungry family food is of far greater importance than shelter, especially where the climate is dry and warm . . . The needs of the poor, or at least the ordering of their priorities, are frequently misunderstood by professionals' (Gilbert, 1992, p. 115). It is quite wrong, however, to assume that the poor do not care about their environment simply because other issues are so pressing. There is evidence that the poor are concerned about sanitation in urban settlements and perceive more space as highly desirable. For example in Kiandutu, a low-income settlement in Thika in Kenya, residents who commented on the problems they faced there highlighted how bad and dirty it was and that there were no toilets or latrines. In another Thika settlement, Makongeni, the non-repair of toilets was highlighted (Andreasen, 1989). In another example, project managers in Monrovia, Liberia were rather surprised to find that residents were quite prepared to pay for certain improved sanitation facilities and to work for a more sanitary environment (Stephens, 1991). In Kenya, a competition which elicited information on children's attitudes to traditional housing found that its spaciousness was viewed very positively, with specific comments made about not needing to 'crowd' into one room for sleeping and having separate rooms for cooking (Lee-Smith and Chaudhry, 1990). Also while the emphasis on the health aspects of overcrowding is perhaps understandable, there are psychosocial problems associated with overcrowding, and in many cases cultural norms about gender and age mixing cannot be respected. This aspect of overcrowding is rarely discussed, although there is brief mention of the issue in Hardoy and Satterthwaite (1992). The phenomenon of households living in a single room (often rented) is extremely common throughout urban Africa. The nature of the household, however, must influence how problematic this situation might be. Few difficulties should arise when the household consists of a couple alone or with young children, but sharing with older children or other members of an extended family may be much more difficult. In some cases room space is shared with other, unrelated households and there can be little doubt that this must be stressful if the households consist of families rather than single people.

Environmental controls and responses in Harare

Harare is the capital city of Zimbabwe and had a population of around one million by the late 1980s. It is not therefore a large city by world standards, and does not suffer from major environmental problems due to traffic congestion or industrial pollution. Prior to independence in 1980 its physical development was extremely closely controlled, and in-migration by Africans was restricted by a variety of legislative measures (Mutambirwa and Potts, 1990). The African urban population was housed in segregated townships (now called high-density areas (HDAs)) and hostels.

Township houses were almost exclusively rented and formally built of modern materials. All had water connections and flush toilets, either connected to the main sewerage system or to septic tanks – and there were no major sanitation problems. There were many lodgers in the rented family housing, but even this housing sector was controlled. Although many families took in unregistered lodgers, it is difficult to know to what extent this caused overcrowding.

Since independence Harare has grown rapidly as influx controls were lifted. A major urban housing shortage developed, estimated at 240,000 units nationally by 1985, as government policy on low-income housing was unable to deliver the numbers of dwelling units needed. The main strategy chosen for the low-income sector was site-and-services, but not nearly enough schemes have been funded to meet demand. The government was quite unable to meet the housing investment plans contained in its first development plan, which stated that the urban housing backlog could be eliminated in eight years, as the economy did not perform as strongly as was hoped. Furthermore the nature of Zimbabwe's site-and-service schemes has severely constrained the rate at which housing units for single-family occupation can be completed. European-style modern building standards, inherited from the previous white-minority government, were maintained and strictly enforced. In theory scheme beneficiaries are meant to construct the minimum of a four-room, rather large house, with ablution facilities within eighteen months. Regular building inspections ensure that standards are adhered to such as poured concrete floors, and fired brick or cement block walls, glass windows with modern frames, and asbestos, metal or tile roofs. The expense of such building materials and construction, exacerbated by a serious shortage of building materials such as cement in the later 1980s, has meant that many people simply cannot afford any of the government's low-income housing options (see Potts with Mutambirwa, 1991).

In many other African countries to the north of Zimbabwe such a shortage in legal housing options would undoubtedly have led to unplanned settlements appearing where more space and lower building or rental costs would pertain. However both city and national government resistance to this has been absolutely implacable in Zimbabwe, just as it was with the previous settler government. The national government's commitment to modern, high-standard dwellings was partly ideological, but now seems more influenced by concerns over image, rather than the people's needs. In the city government, some pre-independence personnel with inflexible attitudes to unplanned housing are still involved in planning, and in many cases even the newer personnel have received training more appropriate to First World cities. Since there is no land under communal tenure adjacent to the city boundaries, let alone within them, unplanned settlements with legal rights to land cannot emerge in Harare, and any squatters on private and even public land receive short shrift. A recent example was the removal of a squatter settlement on public land in Mbare HDA in 1991, just before the Commonwealth Heads of Government meeting in the city. As a result of this severe anti-squatting stance, which is probably the most stringently enforced in Africa, environmental problems associated with unplanned settlements (for example irregular and temporary building materials, very small, poorly constructed dwellings, lack of roads, water and sanitation) have not emerged in Harare. The housing shortage has nevertheless created less visible, but significant, environmental problems.

The shortage of site-and-service schemes, the fact that only about one-fifth of low-income households are estimated to be able to afford to construct the minimum legal house (ibid.) and the stringent controls on unplanned housing mean that the urban population has had to crowd into the existing legal housing stock. Although the government's official policy is for home ownership, in fact the one household per rented room phenomenon is extremely common. These are often lodgers in houses also occupied by the owners (78 per cent of family housing units in Harare were held under home-ownership agreements by 1985). However, the extremely high demand for shelter and the consequent high rents mean that some of the newer houses in the site-and-service schemes are developed, in contravention of the ethos and the rules of the schemes, by absentee landlords. Although they cannot take the form, so common in West and East Africa, of rows of rented rooms, the function of such houses as investments for rental profit is exactly the same. Commonly, each room in the house will be rented out separately. Other site-and-service beneficiaries live in their houses but rent out rooms to help pay for the mortgage and construction. Seriously overcrowded conditions within both the old townships and the newer site-and-service schemes are therefore very common – some indicative data are shown in Table 12.1. This table also shows that multiple household occupancy is not the only cause of overcrowding – the oldest HDAs such as Mbare tend to have very large primary households, presumably because the owners' adult children find it impossible to find alternative accommodation and cannot move out.

Another form of housing which has developed since independence is the backyard shack. The former township houses all had small yards, and in these spaces small dwellings have been constructed from a huge variety of building materials. These shacks are Harare's equivalent of shanty dwellings elsewhere, but they are mixed in with the formal, modern housing and do therefore have access to the services provided for the townships (for example water, sanitation, transport, schools). However, the multiple occupancy of the primary houses already puts a strain on sanitation facilities and the backyard shacks add further to the problem. The urban infrastructure was not designed for such heavy demands; in fact the rapid

Table 12.1 Overcrowding in the Harare high-density areas

High-density area	Survey date	Average number of persons per room	Percentage of all households sharing a dwelling	Average total occupancy rate of houses
Townships:				
Mbare	1982	2.4	30	na
Mabvuku	1982	2.0	31	na
Highfield	1982	1.9	61	na
Dzivarasekwa	1982	1.8	47	na
Kambuzuma	1982	1.7	87	na
Hostels	1982	4.2–5.7[a]	na	na
Site-&-service schemes:				
Glen View	1981	na	na	17–18[b]
Chitungwiza	1989	na	na	17[c]
All HDAs	1983	na	na	6–8

Notes: [a]The range is for singles' hostels and family hostels, the latter having been converted to family occupation.

[b]Glen View houses tend to have a large number of rooms, averaging 6.8 in 1982. This 1981 survey included people who were sometimes absent from the city; another survey in 1982 gave an average *de facto* occupancy rate of fifteen.

[c]Estimate assumed to include all 'stand' occupants, i.e. it includes lodgers not in main house.

Source: Potts with Mutambirwa, 1991

growth of the population of Chitungwiza, Harare's dormitory suburb, was in part indicated in the 1980s by the evidence of the overloading of the sewers due to the overcrowding on housing plots.

The psychosocial stresses caused by such overcrowding are significant and were highlighted in a newspaper report in 1990 (*Herald*, 1990). Examples of socially problematic overcrowding included the case of a man, his wife and five children renting a room not much bigger than a double bed. A piece of cloth screened off the living area from a passage to another inner room, where another family lived. Another example was a woman in Masvingo, in central Zimbabwe, who had to share a room with her husband, adult daughter and grandchild. In the same town there was a case of three families living together, totalling thirty-six people, where the children had to sleep under the beds. In Dzivarasekwa HDA in Harare, a couple shared their room with three daughters, one of whom was fifteen years old. In Chitungwiza a man had his six children and his brother-in-law and his three children all in a two-room house. In another case, a couple in Glen View in Harare with seven children had two rooms. Apart from the sheer physical discomfort involved, and the physical health risks, the report specifically addressed the moral and cultural taboos which were being broken and which were clearly of concern to the people involved. Sharing sleeping space with post-pubertal children was obviously a problem, nor was it desirable for older children of different sexes to sleep in the same room. In some cases the families kept the genders separate by the husband sleeping with the boys, and the wife with the girls. In one of the cases cited above, the seven children all slept together; and this was a problem because the oldest boy was eighteen and the oldest girl fourteen. In another case a man had to move out to a boarding house when his mother-in-law came to stay with two more children. It was also felt that such overcrowding led to children losing respect for their parents and that sharing washing facilities with adults from other families might expose children to moral dangers. In typical journalistic vein the report stated that these conditions 'threatened the country's social fabric', and while such a conclusion is dubious, it is clear that social norms are being flouted, leading to mental stress.

Overcrowding might be eased if lower building and infrastructural standards were allowed, decreasing the cost of the servicing of sites and of house building, thus allowing more schemes and houses to be completed. However, there is now a great emphasis in Zimbabwe on involving the private sector, and particularly building societies, in the provision of 'low-income' housing (Potts with Mutambirwa, 1991). This shift creates yet another barrier to a more rapid and extensive housing programme, since the building societies would not lend for properties which do not comply with expensive, modern standards.

Paradoxically, despite the high standards enforced in the site-and-service schemes where flush toilets have to be installed in each house, there are still sanitation problems. On the one hand, poor beneficiaries find it very difficult to complete their houses because of the inappropriate standards and sometimes live in temporary 'shacks' on the plot for years, with no or very inadequate sanitation. In one of the earliest schemes, Kuwadzana, people were not allowed to live on their sites until they had built one room, a toilet and a kitchen, but this merely hindered completion rates even more since beneficiaries had then found it difficult to work on, or supervise the building of, their houses. Another problem stems from the growing landlordism – for sometimes absentee landlords disconnect toilets, which they had to install to meet the building inspections, and both resident and absentee landlords may restrict lodgers' access to water. This is because the water is metered, and water bills are the landlords' responsibility. The lodgers may then be in the paradoxical situation of being surrounded by modern sanitation infrastructure but actually unable to dispose of body wastes in a sanitary fashion. Even pit latrines would be discouraged by the landlords since they would draw the inspectorate's attention to the breach of building regulations.

On a more positive note, the one unplanned settlement associated with Harare which has survived has benefited from the excellent low-cost water and sanitation technology for which Zimbabwe is justly renowned. This settlement, Epworth, has a population of over 35,000, and has survived because it was on mission land outside the city boundaries (the closest Harare residents can get to settling on peri-urban communal land as elsewhere in Africa). An upgrading programme began in 1985 and has generally been successful. At that time it was estimated that 63 per cent of residents were using unprotected wells and 8 per cent unprotected stream water, while 94 per cent had unimproved pit latrines (*Herald*, 1986). By 1988 safe water sources had been provided for most residents, and many Blair toilets (ventilated improved pit latrines or VIPs) had been built with government assistance (*Herald*, 1988).

Zimbabwe has a good record of research on, and implementation of, low-cost technologies appropriate for low-income groups. If these were allowed to be applied in urban settlements, they might help to improve the urban environment for the poor, but at present the insistence on 'high standards' and strict planning controls is in fact helping to create some environmental problems.

Urban environmental issues in Lilongwe

Lilongwe was designated as Malawi's new capital city in 1975. Its close association with the Life President, Hastings Banda, meant that the nature of its development was tied up with his prestige, and aesthetic considerations nearly always took priority over the needs of the low-income population (Potts, 1985). Lilongwe has been one of the most closely planned and controlled urban developments in Africa, but like Harare, this has not prevented environmental problems from emerging.

As in Harare, major problems of overcrowding exist in the low-income site-and-service settlements, which are known in Malawi as traditional housing areas (THAs) and are the main official housing options for the poor. There are also associated problems with sanitation. The causes of these problems are rather different from those in Harare, although the underlying problem of housing shortage is the same.

In complete contrast to Harare, Malawi opted for very low-cost, low-technology building standards on its site-and-service schemes. Initially this pragmatic approach, which for example allowed the use of sun-dried bricks and beaten earth floors, but not wattle and daub, in construction, allowed the poor a real chance to take advantage of the schemes and 'might be considered in some respects a model for such developments' (Pennant, 1990, p. 197). Unfortunately the planning authority responsible for the THAs was not really committed to low-income settlements, being a market-orientated development corporation, and as in Zimbabwe, the rate of scheme development did not match demand – in fact no new THAs were developed after 1982 (Potts, 1989). The schemes tended to be located far from the town centre, and even from main roads, creating major transport problems for residents. The 1977 destruction of a major scheme containing nearly one-quarter of all serviced plots because it was too visible, and therefore conflicted with Banda's desire to maintain Lilongwe's garden city image, and the reallocation of its 5,970 people to other more peripheral schemes (Pennant, 1990), exacerbated a critical housing shortage.

As in Harare, the housing shortage was rapidly translated into a burgeoning rental sector and serious overcrowding. Between 1972 and 1980 the average number of people per plot doubled to 12.5, and over 50 per cent of the low-income population was estimated to live on plots with occupancy rates of over fourteen. By 1982, in one area, with good access to employment, a third of residents lived on plots with more than twenty-one other people. Most plots were nearly completely taken up with rental outbuildings and extensions to the main house. These overcrowded environments occurred in a situation where there was absolutely no shortage of land for development, since the Capital City Development Corporation had been given ownership and control of 67,842 acres, an area vastly larger than the city itself. The overcrowding problem also reflected the government's resistance to squatting which, at least until recently in Lilongwe, was as fierce as in Zimbabwe. However, here peri-urban land under communal tenure, or under the control of small landowners in the Lilongwe Land Development Programme, has been utilised by some of the low-income population for housing, although such settlements are rather remote from employment areas.

In the site-and-service schemes water is provided from kiosks at one per hundred plots, and sanitation is provided by pit latrines. These low-cost solutions can yield safe and satisfactory environmental conditions, even in urban situations. However, there are far more households per kiosk than planned, increasing queuing time for water, and although the latrines were expected to have a life of seven years, they tend to be full in half that time. Each plot was planned with space for two latrines, but in one area the latrines were already full by 1984. Sanitation is also a problem in the few genuine squatter areas that have emerged within the boundaries, when evictions failed to keep up with their development. However when a new zoning scheme was developed in 1986, only one of these settlements was to be upgraded, and the rest were earmarked for demolition.

Housing and environment in Gaborone

Gaborone is another tightly planned city. Like Lilongwe, it has been developed since independence as Botswana's capital city because during the colonial era the country was administered from Mafeking in South Africa. Thus there was the opportunity to control the development of the urban environment, rather than react to organic and unplanned growth as is the case in most African cities.

The planning response to the needs of the low-income population in Gaborone has differed from that in Harare or Lilongwe and has been rather more successful in delivering a reasonable urban environment than either. The context of the planning was markedly different. In Harare, the government had to contend with a legacy of racial segregation, peripheral townships and an entrenched code of building standards. In Lilongwe, the master plan was drawn up by a South African firm and reflected its segregationist tendencies and the Life President's rather inappropriate commitment to a garden city image. Although colonial Bechuanaland had been strongly influenced by South Africa, the development of a brand new city gave the country the opportunity to try out new, non-segregationist approaches, and it avoided an over-commitment either to aesthetic considerations or high-tech approaches.

The Botswanan government commissioned a plan from the Swedes, who adopted a far more egalitarian approach than is common in southern Africa. Yet again a site-and-service approach was adopted for low-income groups, but in direct contrast to Harare and Lilongwe these schemes have been closely integrated into the city and are not condemned to the periphery. High- and middle-income residential areas can be found absolutely adjacent to the schemes. As in Lilongwe, low construction standards have been allowed, and the initial provision of site infrastructure is very low. Water is provided by standpipes, there is no electricity and no sewage system – instead the residents must build pit latrines. All these measures help to keep the schemes affordable for the poor, since development and service charges will be low, and they can choose the construction materials which are within their means. The level of overcrowding found in Harare and Lilongwe has not appeared, and many plotholders are consolidating their residences to fairly high standards; in general the environmental conditions in these schemes are very reasonable. Observation of some of the schemes in 1991 suggested that one improvement that could be made would be to encourage residents to build latrines which include the crucial elements of VIPs, which help to prevent contamination of food etc. by faecal material spread by insect vectors. These are very simple: it is necessary to keep the inside of the latrine fairly dark (for example by *not* having windows and having the entrance offset in a spiral arrangement) and to place a piece of wire gauze over the vent pipe. These extremely cheap techniques ensure that flies which enter the pit are attracted by the main light source, the light at the top of the pipe, and are trapped and die. Most of the latrines seen did not have this gauze, and many had windows.

During the early construction period in Gaborone when housing policy for the urban poor had not been fully developed, a squatter settlement called Old Naledi developed. In 1977 this consisted of about 10,000 people on 2,000 plots. Environmental conditions were poor: about a quarter of the structures were of temporary materials, the majority of people used fields to the east and west of the settlement to defecate – there were only about a hundred simple pit latrines in the whole area – and there were only four taps for the entire population, all in one corner of the settlement. Many people therefore had to walk one or two kilometres for water. The government at first tried to get rid of this community; site-and-service schemes which allowed the use of traditional construction methods were set up with the direct intent of enticing the squatters to relocate, and then a contractor-built scheme of small, mass-produced houses was developed to which the squatters were to be moved. Neither policy was successful, and in 1977 the government decided to upgrade Old Naledi and give the people security of tenure. This programme was completed with the approval and involvement of the community. Standpipes were provided throughout the settlement, with provision for individual metered connections to private plots if these were paid for. All plots had to have room for a pit latrine in order to get security of tenure, and the sub-structure of a double VIP was provided for each plot (allowing one side to be emptied, while residents can use the other side). Thus environmental conditions in Old Naledi were greatly improved, at a reasonable cost, and with the support of the residents.

The Botswanan example does contain lessons for other southern African urban centres, but it should be noted that this country does not have to contend with the financial shortages that the rest of the region faces. Botswana has had one of the fastest-growing economies in the world during the past two decades, based on its diamond wealth. It has a hard currency, so shortages of construction materials and escalating costs have not been a major problem. The low-income urban population probably has relatively higher incomes than the poor in, for example, Harare and their incomes would be more secure since unemployment is very low. None of these circumstances, however, makes a pragmatic planning approach any more likely. Botswana's 1975–81 development plan contained a number of urban development policies which suggest the government was trying to avoid some of the problems which had arisen elsewhere in Africa and which were beginning to emerge in Botswana in the early years of independence. These policies included a commitment to try to restrict subsidisation of urban development by ensuring that standards of urban infrastructure can be paid for by residents, although cross-subsidisation of low-income housing areas from the sale of high- and medium-cost plots was to be allowed. The government also wished to prevent economic or social polarisation in order to preserve the social health of urban areas. A further aim was to provide an adequate supply of land serviced at standards affordable even to the poorest members of the urban community, in order to prevent speculation (van Nostrand, 1982).

Informal settlements in Durban and Johannesburg

The problems faced by the low-income housing sector in South Africa are on a rather different scale from those discussed in the examples above. It is estimated that there is a backlog of one million houses in the urban areas, for example, and the growth of informal settlements has been absolutely explosive in the past ten years or so, as political conditions have undergone major changes. The issues involved in low-income housing here are very complex, given this country's unique attempt to control every aspect of urban settlement until recently, and much has already been written about contemporary urban problems (for example Tomlinson, 1990; Swilling *et al.*, 1991). The following discussion will briefly introduce some urban environmental issues which relate to those already raised in the previous southern African examples.

Durban probably has the greatest proportion of its low-income population living in unplanned housing. In 1991 it was estimated that the greater Durban area had a population of about four million, of which 50 per cent were in informal settlements. About 65 per cent of the African population was believed to be housed in such settlements – a complete transformation from the idea of the apartheid city, where all housing was meant to be closely planned and controlled. These informal settlements form three main categories. The phenomenon of backyard shacks, which have become so common in Harare, is very prevalent in South Africa. The township housing and plots have many common features in these two countries, and the severe housing shortage in major metropolitan areas caused by government policy led to the development of unplanned secondary dwelling units on the legal rented plots. These are often rented out or may house extended family members of the main tenant. The second category of informal settlement is what are sometimes called free-standing shack areas (FSSAs), in order to distinguish them from backyard shacks, but which elsewhere would be known simply as squatter settlements. In Durban, however, many of these are not strictly squatter, being on peri-urban land under the control of KwaZulu homeland leaders. The final category of informal settlements is site-and-service schemes, which are denoted 'informal' in South Africa, despite the fact that they are planned and legal. Many of these are being developed, but as yet they house only a minority of those low-income residents in 'informal' settlements.

The environmental standards in the FSSAs are very variable. A key variable in South Africa appears to be the level of political organisation in such settlements. The low-income urban population of South Africa must be one of the most politicised in the world, and many settlements have very strong civic associations. In some cases there have been planned invasions of land, in a similar fashion to those in urban Latin America – nothing like this occurs in the other countries under consideration. In these cases there must be some degree of political organisation. Often this leads to a fairly well laid out settlement, with defined roads and plot boundaries, and the association is able to negotiate for some services. However, although the scale of FSSAs in South African cities now makes it quite impossible for the government to demolish all of them, they are individually still very insecure, and many local governments are strongly opposed to this form of housing. Most cities do not have peri-urban homelands as Durban does, so even peripheral FSSAs are squatters. Insecurity of course discourages residents from improving their environment by investing in permanent housing materials and larger dwellings or adequate pit latrines. However, in South Africa insecurity does not only stem from tenure status, since the level of political violence in many settlements is extreme. The settlements around Durban, for example, have seen terrible destruction in conflicts between Inkatha and the ANC.

One of the largest unplanned settlements is Inanda in Durban. Sanitation has been a major problem. The standards of construction of pit latrines in the area are very low indeed and would not greatly improve hygiene as VIPs would. At one point, when the Umgeni River was found to be polluted, these latrines were blamed for the contamination, but research in the area found that this was not the case. In fact, people were rarely using the latrines at all, probably because they were so dirty, and sanitation conditions must be a major health hazard. Overcrowding is also rife and this must affect disease transmission. Death rates of 26 per cent from measles in babies aged under eight months have been recorded for urban Natal (Pryer and Crook, 1988), and conditions in Inanda would certainly facilitate such appalling mortality.

Some of the FSSAs on unused land in Soweto in Johannesburg have truly desperate environmental conditions, with no sanitation, plastic sheet housing and dreadful overcrowding. By contrast, planned invasions such as Stanzaville in Mamelodi township in Pretoria seem well ordered and clean. Here the civic association laid out plots for residents before construction began. These are, however, very small, being only 80 square metres. The community numbered around 4,000 households by August 1991, a year after the initial invasion. Most of the houses were made of packing cases with zinc roofs and probably provided just about a minimum standard of reasonable shelter. However, it has to be remembered that Peil's (1976) assertion, about the lesser significance of shelter standards in tropical conditions, does not hold in the Transvaal in South Africa where the winters can be very cold. Sanitation in Stanzaville was provided by chemical toilets on each street, for which each household had to pay R5 per month. This was felt to be far in excess of cost recovery and was much resented. Water was a major problem for the whole community. Improved services were gradually being phased in, but by August 1991 there were only a total of ten taps in the first phase and none elsewhere.

The backyard shack phenomenon, as in Harare, is mainly characterised by environmental problems caused by overcrowding and overuse of the toilets which were designed for use by one household. The Urban Foundation estimates that 80 per cent of the population of informal settlements live in backyard shacks in the Pretoria–Witwatersrand–Vereeniging metropolitan area. The development of FSSAs to some extent reflected the need to relieve the desperate overcrowding and social stresses associated with the backyard shacks, but it is doubtful that average population densities on the formal township plots have fallen, since other desperate households would probably soon have taken the place of those who moved out.

Site-and-service developments in South Africa have, perhaps unexpectedly, avoided the problem of insisting on the over-high and inappropriate building standards which cause problems in Harare. In general residents can use almost any construction materials, and the availability of packaging products from major manufacturing firms in South Africa means that these are frequently used. However, the standard of sanitation facilities tends to be very low which is somewhat surprising since it might be expected that South African planning authorities should have the capacity to implement decent pit latrine technology with community involvement and government assistance, as in Old Naledi in Gaborone and Epworth in Harare. In fact many of the so-called site-and-service schemes are misnamed because no services are provided at all, and arguably these schemes suffer from being underdeveloped. Cost-cutting is so extreme that many residents refuse to pay their service charges on the grounds that there is nothing to pay for. An example of a minimalist scheme is Orange Farm, built for Johannesburg low-income households. Here there is one tap per twenty households, and residents must build a pit latrine on their small stands. The latrines are generally of very low standard. Perhaps the most extraordinary feature of Orange Farm is its location – it is sixty kilometres from Johannesburg on farm land. Thus not only do residents have rather minimal housing conditions, but extra stress is created by the problems over transport to work. It is therefore not surprising that many refuse to pay the R29 per month service charge. Stand costs were originally R500, but in 1991 it was planned to increase this to R2,000–R3,000. It is hard to see how this could be justified.

Conclusions

The very severe economic crises which have typified most African economies during the 1980s have had a serious impact on urban development and the capacity of urban management to maintain or improve the urban environment. In many cities environmental conditions in low-income settlements have worsened, as economic problems have taken their toll on urban water, sanitation, low-income housing and transport programmes. There is evidence of worsening malnutrition and child mortality rates in some cities. In low-income settlements serious health and social problems arise from overcrowding on plots, within houses and from poor sanitation.

Although many governments face serious constraints, the evidence from the southern African cities discussed in this chapter shows that planning responses to low-income housing needs can still make a difference to the environmental problems faced by the poor. Poor people feel strongly about the need for a clean environment (Sinnatamby, 1990), and despite their different priorities in terms of space, it is clear that certain types of overcrowding are highly stressful. The standards implemented in low-income housing projects are important determinants of environmental conditions: too high and overcrowding may be encouraged; and if they are too low, then few improvements in people's physical and mental health may be achieved. There are also low-technology, yet adequate, sanitation techniques, but these do require some planning input at least in terms of education. In each of the four countries considered in this chapter, there have been both positive and negative aspects to the environmental impact of the planning response to low-income housing needs which suggests that these countries have much to learn from one another.

References

Andreasen, J., 1989, 'The poor don't squat: the case of Thika, Kenya', *Environment and Urbanization*, **1**(2), pp. 16–26.

Ayeni, B., 1981, 'Lagos' in M. Pacione (ed.), *Problems and Planning in Third World Cities*, London, Croom Helm, pp. 127–55.

Bradley, D., Stephens, C., Harpham, T. and Cairncross, S., 1992, *A Review of Environmental Health Impacts in Developing City Countries*, Discussion paper 6, Urban Management Program, Washington, DC, UNDP/World Bank/UNHCS, World Bank.

Gilbert, A., 1992, 'Housing of the urban poor', in A. Gilbert and J. Gugler (eds), *Cities, Poverty and Development: Urbanization in the Third World*, Oxford, Oxford University Press.

Hardoy, J.E. and Satterthwaite, D., 1992, *Environmental problems in Third World Cities: An Agenda for the Poor and the Planet*, London, International Institute for Environment and Development.

Herald, 1986, 'Rags to riches as Epworth gets a facelift', Harare, 6 January.

—, 1988, 'Epworth projects near completion', Harare, 5 June.

—, 1990, 'Overcrowding, lack of adequate housing threaten country's social fabric', Harare, 8 August.

Izeogu, C.V., 1989, 'Urban development and the environment in Port Harcourt', *Environment and Urbanization*, **1**(1), pp. 59–68.

Kulaba, S., 1989, 'Local government and the management of urban services in Tanzania', in R. Stren and R. White (eds), *African Cities in Crisis: Managing Rapid Urban Growth*, Boulder, San Francisco and London, Westview Press, pp. 203–45.

Lee-Smith, D. and Chaudhry, T., 1990, 'Environmental information for and from children', *Environment and Urbanization*, **2**(2), pp. 27–32.

Mutambirwa, C.C. and Potts, D., 1990, 'Changing patterns of African rural–urban migration and urbanisation in Zimbabwe', *Eastern and Southern African Geographical Journal*, **1**(1), pp. 26–39.

Novick, R., 1976, 'The relationship of the environments of human settlements to health', in P. Laconte (ed.), *The Environment of Human Settlements: Human Well-being in Cities*, Volume 2, Oxford, Pergamon Press, pp. 30–42.

Omer, M.I.A., 1990, 'Child health in the spontaneous settlements around Khartoum', *Environment and Urbanization*, **2**(2), pp. 65–70.

Peil, M., 1976, 'African squatter settlements', *Urban Studies*, **13**, pp. 155–66.

Pennant, T., 1990, 'The growth of small-scale renting in low-income urban housing in Malawi', in P. Amis and P. Lloyd (eds), *Housing Africa's Urban Poor*, Manchester and New York, Manchester University Press.

Potts, D., 1985, 'Capital relocation in Africa: the case of Lilongwe in Malawi', *Geographical Journal*, **151**(2), pp. 182–96.

—, 1989, 'Urban environmental control in Southern Africa with special reference to Lilongwe', *Resource Management and Optimization*, **6**(4), pp. 321–34.

Potts, D. with Mutambirwa, C.C., 1991, 'High-density housing in Harare: commodification and overcrowding', *Third World Planning Review*, **13**(1), pp. 1–26.

Pryer, J. and Crook, N., 1988, *Cities of Hunger: Urban Malnutrition in Developing Countries*, Oxford, Oxfam.

Sinnatamby, G., 1990, 'Los cost sanitation', in J.E. Hardoy, S. Cairncross and D. Satterthwaite, *The Poor Die Young: Housing and Health in Third World Cities*, London, Earthscan, pp. 127–57.

Stephens, C., 1991, 'Back to basics: a community based environmental health project in West Point, Monrovia, Liberia', *Environment and Urbanization*, **3**(1), pp. 140–6.

Stren, R. and White, R. (eds), 1989, *African Cities in Crisis: Managing Rapid Urban Growth*, Boulder, San Francisco and London, Westview Press.

Swilling, M., Humphries, R. and Shubane, K. (eds), 1991, *Apartheid City in Transition*, Cape Town, Oxford University Press.

Tomlinson, R., 1990, *Urbanization in Post-apartheid South Africa*, London, Unwin Hyman.

van Nostrand, J., 1982, *Old Naledi – the Village Becomes a Town: an Outline of the Old Naledi Squatter Upgrading Project, Gaborone, Botswana*, Toronto, James Lorimer.

World Health Organization, 1988, *Urbanization and its Implications for Child Health: Potential for Action*, Geneva, WHO.

13 Traditional housing and future urban planning strategy in Katsina, Nigeria

Michael Mutter

For a period of ten years, from the mid-1970s, I was fortunate to be a member of the small team of multi-disciplinary urban planners formed by Max Lock to undertake urban development studies in northern Nigeria. An intrinsic part of the work involved the study of the existing traditional housing stock as physical and socio-cultural factors in the urban dynamics of the region. We were able to come to conclusions about the positive role for traditional housing, especially in the process of rapid urbanisation, based on the experience of people in these towns. In particular, in 1977–8 Max Lock Group Nigeria was commissioned by the then Kaduna State Government to undertake surveys and prepare development plans for the seven local government headquarter towns in the northern part of the State. I directed the study and edited the final report (Mutter, 1978). Katsina, the largest of these towns and regional capital, provides a useful case study for examining the role of traditional housing in the urbanisation process.

Katsina: the Hausa city

Katsina is an ancient walled city in the extreme north of Nigeria. It is one of the original Hausa Bakwai, the legendary seven towns that formed the home of the Hausa-speaking people hundreds of years ago. An indigenous Islamic city, Katsina gained its pre-colonial influence from the north with trading routes across the Sahara. It continues to be the seat of an Emir commanding an Emirate region across some 2,000 square kilometres. The British occupied Katsina in 1904, forming a province in 1934 as part of Lugard's dual-mandate era. Katsina is now a state capital and a major government centre.

The town is defined by the remains of its encircling walls, huge defensive earth ramparts that once enclosed not only the built-up settlement but also a considerable extent of farmland, so that it was able to withstand siege. Both gained water from shallow wells within the walls. Nine gateways controlled access to the town. Outside the walls, farmers made use of the seasonal watercourses to develop *fadama* (irrigated and intensively productive vegetable gardens). The town itself derives its character from the densely built-up agglomeration of mud-walled compounds, traditional Hausa houses; much of it is quite old, but new parts are replicated quite naturally in the same pattern.

Katsina is now growing at a dramatic rate. In 1977, when it was surveyed, the town's population was growing annually at 4.69 per cent, based on an estimated population of just over 100,000. As a state capital since 1988, Katsina's rate of growth will have increased. Even at an annual 5 per cent growth, the expected population by the year 2000 would be 310,000 and the new residential area requirement would be 2033 hectares, more than four times the 1978 residential area (Mutter, 1978). For housing alone this would exercise a current demand for new land, released and serviced for development, at a rate of 100 hectares per annum. Although this would allow for enhanced standards, it must also compete for suitable provision of access and infrastructure with the requirements of land for commerce and institutions, those often having first priority, all at the expense of the incumbent farmland. A tall order for planning!

Max Lock: urban practitioner

Max Lock was one of the few British professionals in urban development who understood the dynamics of housing growth in the immediate post-war era (Mutter, 1991). He was an architect–planner who, unusually, put great emphasis on the coordination of teams from wide-ranging investigative disciplines – sociologists, economists, geographers, engineers, planners and architects – undertaking a process of what he termed 'civic diagnosis'. This involved in-depth surveys of the physical, social and economic parameters of planning, including household interview surveys. By careful study of the social component, the teams were able to identify the immediate and future natural impetus for growth. The study was married to complementary studies of the physical and economic potential for growth, all to be enshrined in an administrative process that could make it happen. This proved a useful basis for planning, not only in the United Kingdom's reconstruction after 1945 but also as an approach to Third World urban planning.

The Max Lock Group housing philosophy in Nigeria

Max Lock pioneered his approach to planning and urban development in less-developed countries in a way that sponsored local initiatives in planning for the indigenous people. He took the unusual step of putting emphasis on the preparation of detailed household interview surveys, conducted by local people, as a basis for evolving local structure plans. This approach was principally put to work at Kaduna, the capital of the then Regional Government of Northern Nigeria, as a result of an aid programme initiated by the UK Ministry of Overseas Development in 1965. Later, the Max Lock Group Nigeria was formed as a consultancy body in its own right, and was commissioned by various state governments to prepare Development Plans based on the survey techniques that identified trends and directions, needs and capabilities.

The work was concerned with assisting the state governments in planning the towns to cope with the increasing rate of in-migration and growth. New development areas and improvements to the existing urban fabric were identified within wide briefs that included long-term development plans, urban administration, coordination of infrastructure investments, immediate implementation of development layouts and housing area improvements which, from the late 1970s, were in line with World Bank Urban Development Project programmes.

The planning methodology put a great deal of emphasis on socio-economic and physical survey, on procedures for up-grading existing housing areas and on standards and layouts for new self-build housing which had been established by Max Lock in his work on Kaduna in 1965. The publication of the Report (Lock and Theis, 1967) caused a stir among writers concerned with Third World urban development (Blair, 1968; Isaacs *et al.*, 1969).

Towards a housing policy

In arriving at anything like an agreed housing policy in Nigeria, there has been great confusion. Each strand of government initiative has no doubt been well meaning, but because the objectives have been varied in each case, little progress has been made. Yes, the towns have grown rapidly and accommodate many more people; yes, many government-financed houses have been built; and yes, much research has been done to define needs and aspirations; but still there remain widespread overcrowding, insanitary conditions, poor housing environments, many people facing landlord extortion and others without any home.

What, then, is the role that traditional housing can perform to assist in alleviating these conditions in an area such as Katsina State, until 1988 the northern part of Kaduna State, Nigeria? To answer this question means returning to square one and re-evaluating the original product, the Hausa house, and the process by which people can simply build their own homes, individually or collectively, as the accepted housing policy sanctioned by government.

Max Lock Group Nigeria has been able to build up an extensive picture of living conditions and the need for appropriate housing in the towns in which it has worked. This chapter presents a few examples of the research, and on behalf of our Nigerian colleagues, friends and the government officials with whom we have worked, restates our collective view for the future development of housing in the region: a belief in the usefulness of the traditional housing form and, indeed, the simple methods of achieving housing for all.

Traditional housing in Katsina

Katsina is cited by O'Connor (1983, p. 31) as 'an outstanding example' of the urban tradition of the sahel–savanna belt of west Africa. The characteristics of towns in this region are quite distinctive, being defined by a defensive rampart encircling both high-density urban areas and some of the intensive agricultural areas, although in the case of Katsina these latter have now almost all been built over. O'Connor further notes:

> These Islamic cities occupied by mainly Hausa people in northern Nigeria, like those of the Yoruba, show much cultural continuity with the surrounding rural areas, although they have always housed a distinct elite. Much change has occurred during this century, but it has been largely incorporated within the traditional social system rather than producing a very different system as in cities in many other parts of Africa [ibid].

The character of Katsina, like that of other towns in the region, is dominated by the agglomeration of mud-walled compounds, each enclosed by a high wall. These compounds are traditionally subdivided into a series of mostly private spaces, each containing a number of individual rooms. The entrance room, the *zaure*, is usually a male-dominated reception room for visitors and may be preceded by a smaller entrance room to screen the dwelling completely from the public outside. Inside, the compound becomes an increasingly protected series of spaces and buildings occupied by females and children. The public domain outside the compound wall is a hierarchy of receding spaces varying from irregular open areas at intersections of public rights of way, often dominated by a shade tree forming a gathering space, to extremely narrow access footpaths, acting also as the lines of *ad hoc* sullage water drainage and surface water run-off. Sewage is disposed of in pit latrines and often crowded and primitive septic tank systems. More recently, more regular grid-iron street patterns have been developed, but still with the space acting as important gathering and petty trading areas. Several of these features are evident in the Katsina street block represented in Figures 13.1 and 13.2.

In Katsina in 1977 we made a specific survey of living conditions in the urban area. Besides the socio-economic interview survey of approximately one in ten households, typical street blocks were also selected for surveying in greater detail to give an in-depth view of domestic life and general environmental conditions in the residential areas, together with an appraisal of the construction techniques used in building the Hausa house. Ninety per cent of all building types were of mud-walled construction, comprising some 16,000 compounds and covering some 500 hectares. Allowing for vacant buildings (9.7 per cent of the total), we estimated from the sample household survey that there were 21,000 households. There were on average 6.8 people per compound and an overall density of 201 people per hectare. Densities ranged from just under 100 to over 400 people per hectare, and room occupancy rates, although generally well under three persons per room, could in places exceed eight! The density of building and occupation varies widely – from the more traditional low-density compounds still containing much open space and allowing several generations of households to live together, as described in detail by Schwerdtfeger (1982) in his analysis of traditional housing in Zaria, to much higher densities of building, being occupied by many households, often unrelated, and ultimately appearing more like the ubiquitous landlord housing found right across Nigeria.

We found that in spite of somewhat cramped living conditions with poor access and few amenities, people preferred to live in the high-density areas of the towns, close to their relatives and within familiar surroundings. Sometimes, but not always, these areas were also close to places of work. Compared with smaller towns in the Emirate, in Katsina there were not so many people directly farming as a main occupation (9.3 per cent), although many (42.3 per cent) had farming as a secondary and thus seasonal occupation.

Even among the upper strata of the social hierarchy a decision to build a new house 'outside', in other words in the low-density Government Residential Area outside the city walls and away from the traditional family land in the high-density inner areas, was considered with caution. Political and social influence was more easily held and identified by residing inside the old town. Only the politically and financially assured would readily relocate outside and then often for reasons of prestige.

Where new proletarian housing areas were being developed by individual families on customarily acquired land, people were naturally retaining their traditional concept of walled compounds, a single entrance from the street through a *zaure*, with further rooms being arranged around courtyards according to the degree of privacy required for the household structure.

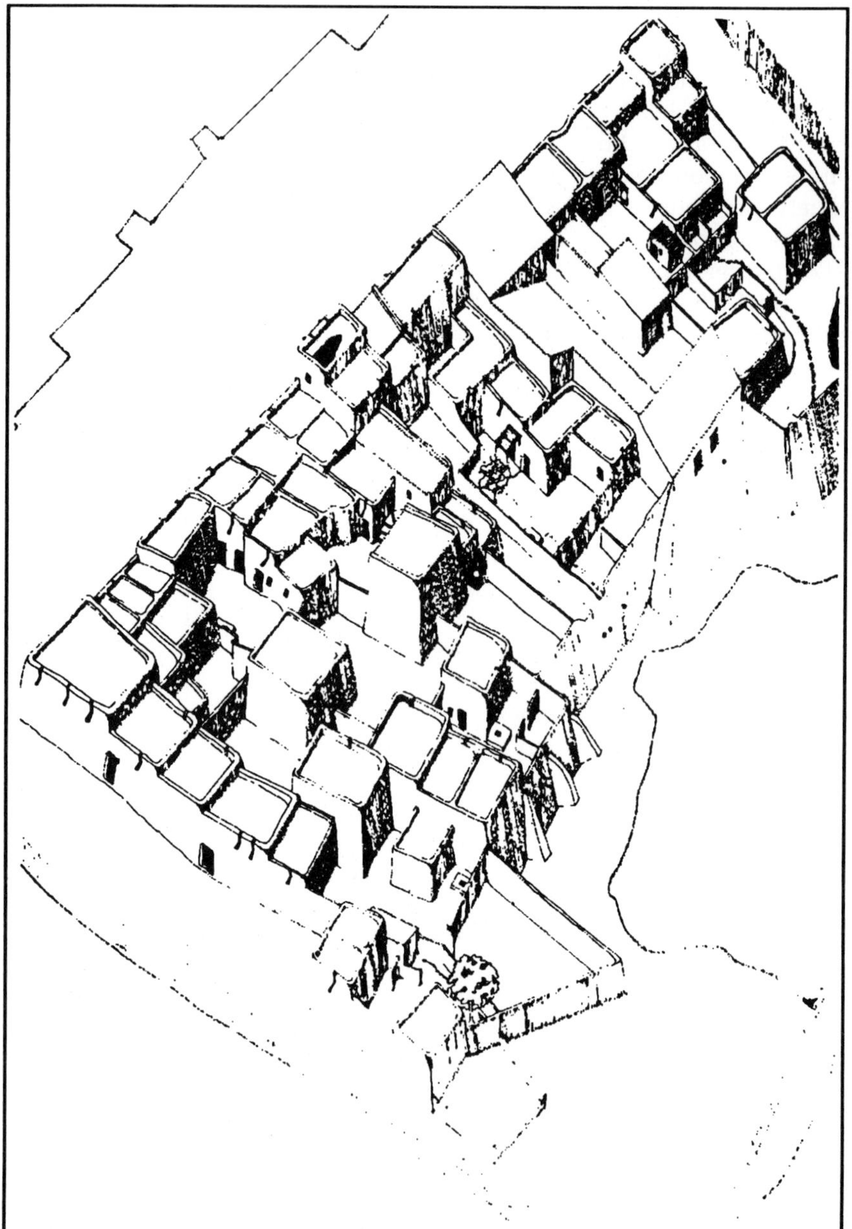

Figure 13.1 Axonometric view of Katsina street block
Source: Mutter, 1978 (Katsina), 2.2.6

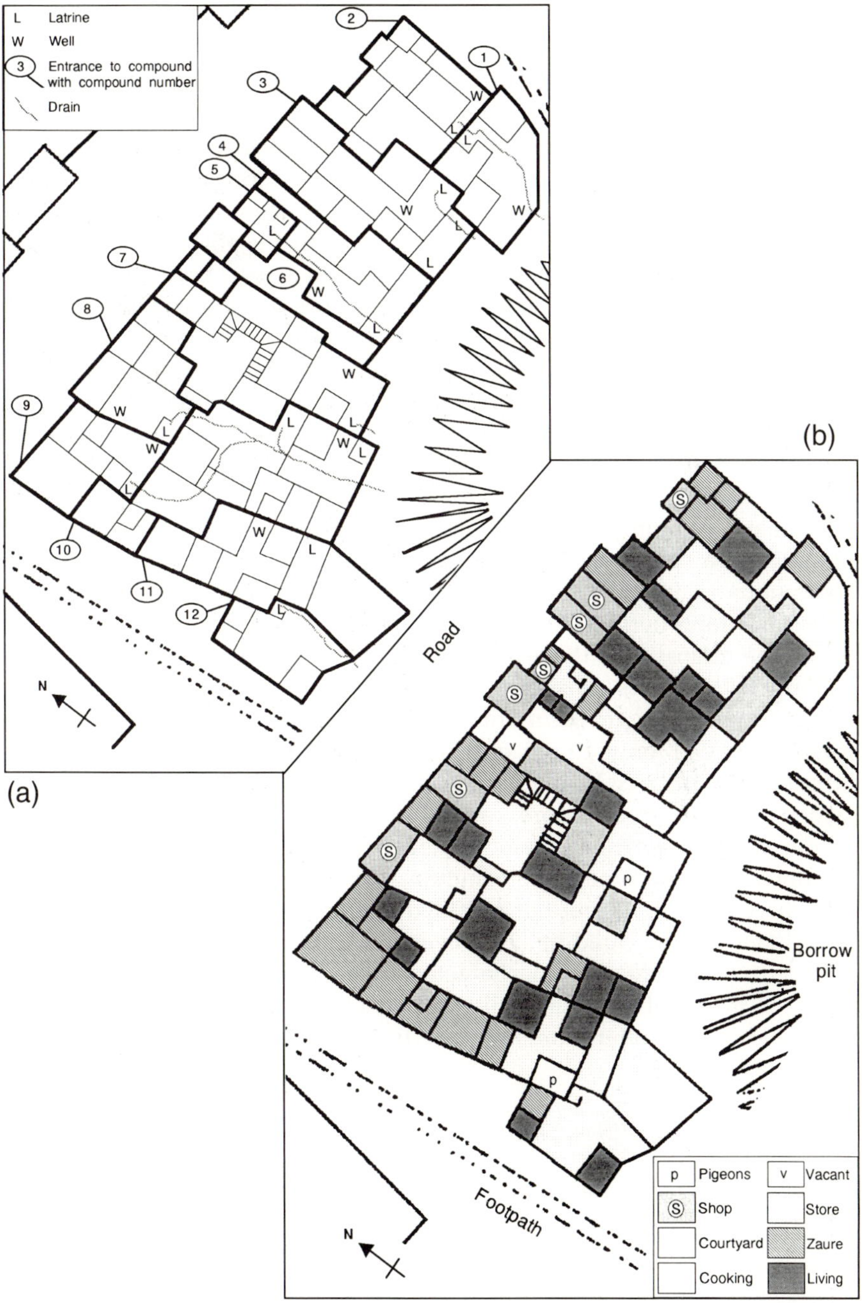

Figure 13.2 Plans of Katsina street block shown in Figure 13.1: (a) Compounds, latrines and wells. Source: Mutter, 1978 (Katsina), 2.2.5 and 5.5.9/10; (b) Land use. Source: Mutter, 1978 (Katsina), 2.2.5 and 5.5.13/14

Traditional building construction

The majority of new buildings were of traditional stabilised-clay construction, a craft skill that considerably exceeds the meagre description of 'mud'. Deftly applied slurry renders are sometimes sealed with a concoction of *makuba* (a local bean sheath) that offers a longer life to the render and a dramatic dark red colour. This colour contrasts with the white weathering cap and clay finials reinforced with the rot- and termite-resistant *azara* (poles split from the tall local palm trees). These are also used for arch construction and for the main structure of mud-covered flat roofs, which characterise the traditional architecture of the region. All this, of course, is achieved without any imported material or high technology.

The basic traditional building material is sun-dried mud cones or mud bricks which are rapidly built up into thick mud-rendered walls and roofed with timber rafters and mud-coated boarding, forming flat roofs and decorated parapets. This form of building is a response to the climate, culture and economy of the region, being readily achievable with a variety of skills on a shared basis, without great investment. However, it requires continuous maintenance and this is related to the pattern of farm labour, which has long periods of relative inactivity in the dry season. Agriculture is still the mainstay of the economy of Katsina Emirate.

Many factors have conspired to change this happy co-existence, however, as the towns have grown more and more apart from the farming link. In Katsina by 1977 only 10 per cent of the population was directly linked to farming. The price of labour is thus much higher. New materials are longer lasting, though much more capital intensive. Thus concrete block and render walling, corrugated zinc-coated steel roof sheeting and glazed steel windows have become more common. They are less ideal climatically but have become the cultural norm, suited to the available investment and life expectancy of the building. However, each new material is treated with sufficient ornamentation to characterise the new idiom. Courtyards, the scene of much of the family life and once swept sand, now tend to be cemented over to offer more intensive use, especially in the rainy season.

Continuity of the traditional housing type

The key to traditional housing is the availability of legally demarcated and allocated land. Given land, traditional Hausa housing thrives on a chance to develop. The availability of land for individual development is complex. It relates to the traditional organisation of local government – originally a hierarchical system administered by the Emir and his representatives.

Thus the problems of a middle- or low-income earner who intends to build his own house start with plot allocation – obtaining the land. All land in northern Nigeria is vested in the state, with a process of allocation of one parcel of land to each citizen as a matter of right, formalised in the Land Use Decree 1978. The problem is that the process and procedure are subject to the methodology of adminstration. This has created both an artificial land market and a waiting list for land allocation.

State governments, being responsible for land matters, have created many housing layouts, which tend to be regular grid layouts of generous plots suitable for high-income group housing. These are reminiscent of the GRAs – Government Reservation Areas – of the colonial era, extremely low density and away from the traditional urban area. In some areas, denser layouts have been created within, or as extensions to, the built-up areas, but they are not prolific and do not meet demand.

The delay inherent in the official plot allocation procedure forces the less privileged to circumvent the process by the non-legal, but customary, purchase of land from local chiefs. They are the informal bottom tier of land control, mainly concerned with local agricultural land, often 'pegged out' into layouts by a local 'surveyor'. Such lands are then built upon without regard for the Town Planning Layouts or District Plans for the area, resulting in haphazard street blocks, poorly controlled and administered, and making the provision of water supply and sanitary services a nightmare. Thus, in spite of state government influence and statutory power over land matters, 'uncontrolled' development springs up, although in reality there is a degree of local as against state government control. These developments, virtually squatter areas, with often considerable personal investments made in the permanent construction of dwellings, lead through inadequate means of access and lack of surface and sullage water drainage to poor environmental conditions and rapidly become the slums of today, let alone the future.

Housing policy options

How can it be that, having put so much effort into the variety of ways of providing housing in Nigeria, millions still live in poor and deteriorating housing conditions? The machineries set up to alleviate the situation have ground to a virtual halt. Yet there is so much will and inventiveness on the part of individual Nigerians to satisfy the demand for shelter for all. Where has it gone adrift?

There are certainly contradictions in the collective administrations governing land, finance and the community. Such contradictions have, no doubt, always existed. But that is no excuse for the future. Future housing policy must surely seek to eliminate the waste of existing skills and traditions in housebuilding and sponsor a good return on the resource of people in adequate housing.

Alternatives tried and the financial implications

Like land, finance plays a major role in the provision of housing and presents a most difficult task in any housing programme. The mass built-housing programme that was embarked upon in the 1970s was an expression of the wealth that was available in that period from oil receipts. State governments, grant-financed by the federal government, itself buoyed up by oil revenues, embarked on a programme of housing units for middle-ranking civil servants. These housing units were built on specific self-contained sites, often inconveniently located for public transport, at medium density, but at great cost for a mediocre specification. They heralded the contract-orientated housing solution, benefiting the builder even more than the occupant. The creation of states in 1967 and more states in 1976 gave an impetus to this kind of housing, all at highly subsidised low rents to relatively well-paid civil servants.

In 1973 the federal government launched the Federal Housing Authority with the aim of housing the middle- and 'low'-income groups but failed to define eligibility precisely. The Authority's immediate responsibility was to initiate, coordinate and execute the construction of 54,000 units as a National Housing Programme. The target was to build 202,000 'low-income' group housing units by 1980. Only a fraction were built. Civil servants, especially the high- and middle-income groups, were given priority as prospective tenants, although some units were allocated by ballot. Again, the journey to work was difficult for many residents from these locations as a rule. Again, rents were highly subsidised. Again, contractors benefited handsomely, even when in Kaduna a block of hastily built flats, part of the programme, collapsed. Design solutions tended to be based on universal multi-storey blocks of flats, easily and quickly built, rather than taking local requirements into account, even when more appropriate schemes had already been drawn for the very sites being developed. Where schemes were able to take account of local requirements and the town planning work that had already been achieved, as at Bauchi, the schemes fell to the inevitable constraint of finance. The targets were impossible to achieve, the results were not part of the local development initiative and so failed.

Although funds were then available from the Federal Mortgage Bank, its loan terms did not attract middle- and low-income earners. Applicants had to have a minimum credit of 500 Naira with the Bank; a 10 per cent down payment had to be made against the value of a house; the Bank offered only 40 per cent of the house value at 6 per cent per annum over twenty years, plus 0.25 per cent service charge, inspectors' fees and other additional charges foreign to traditionally accepted local family house-building and finance arrangements.

Between 1979 and 1983 Housing Corporations were established at state government level. They were charged over these years with taking on the responsibility of the National Housing Programme from the Federal Housing Authority. They were to be essentially profit-making, using a revolving fund. Generally, the State Housing Corporations have continued the FHA policies of contractor-built housing estates and have not financed schemes for lower-cost serviced plots or core units. Thus they have had no effect on the general public housing requirements nor offered any support for the indigenous traditional housing and building abilities.

The provision of even fairly basic infrastructure doesn't come cheap. When coordinated, it can be cheaper to each plot holder than, for example, expensively duplicated water supply pipes that individual households acquire sporadically, each transaction resulting in yet another private water main. On state government layouts the infrastructure has usually been provided as part of the plot. Unofficial layouts rarely achieve a coordinated standard and the question of how to pay for the infrastructure remains. The universal provision of basic infrastructure on new housing developments and urban renewal schemes remains unresolved in terms of individual household accountability and therefore unfulfilled.

Sites-and-services layouts

Until the current plot allocation process is made simpler, faster and readily available to low as well as high income groups, land speculation will continue to increase and the lower income groups will continue to be marginalised by being forced to seek meagre opportunities in poor environmental conditions. In many ways, of course, the sites-and-services programmes of schemes assisted by the World Bank Urban Development Programme and other accountably financed projects have had the same effect. They offer well-planned layouts, a degree of paved access, a good sized plot, water supply, organised surface water drainage, a plan for foul water drainage, electricity supply, some kind of investment plan for a sanitary core and a method of building that is controlled, regulated and available. However, by definition, they tend to be beyond the financial ability of the majority of households because all the costs must be recoverable from the plot holders. This is seen as an injustice since such costs are not otherwise recovered from, for example, the high-cost low-density layouts where plots are allocated to the well off. Until there is some kind of municipal rating system to repay the benefits, providing well-serviced urban land for development at all levels, this imbalance will remain.

To whom the cheapest schemes are available, and to what extent they can answer the demand for affordable shelter in any area, are questions to be answered locally. That they can be used in conjunction with traditional housing is certainly true in Hausaland. In themselves they will stimulate the development of the Hausa house. However, questions remain about whether low-income serviced plots should require the costs to be recovered from recipients, when high-income plots are not. Should not all occupiers of serviced urban land pay an on-going levy to finance a rolling programme of new infrastructure and maintenance investment? And should not this levy be based on income or wealth, or size and quality of investment in property?

An intermediate methodology for the layout of new housing areas

The dilemma of balancing availability of new housing land for individual development with the costs involved was tackled as part of the town-planning work undertaken by Max Lock Group Nigeria. Although initially commissioned by the various state governments, the development plan work also involved the practical aspects of land allocation and especially the processes involved with the local government layouts. These were generally well considered, the land was earmarked and compensation to farmers was organised and perhaps a concept layout was designed on paper; but the stumbling block invariably centred on the cadastral system of land demarcation and the costs involved in the extreme precision of the surveying of property beacons demanded by the system. The delays of method and finance render this approach ineffective.

The answer was to devise an intermediate methodology that would facilitate the same process of plot allocation and security of tenure that a fully fledged local authority layout would have afforded but making use of simpler methods of recording the land area and its limits to development; either physical limits – reasonably flat land free of liability to flood, erosion and rock outcrop – or legal limits. A prime requirement for mapping is recent aerial photography. This is generally the cheapest component in the fairly lengthy and costly process of producing line-drawn mapping. The aerial photography itself can be used for the intermediate layout purposes. This involves using approximate enlargements of the photography, verified for scale on the ground using a simple process of 'dumpy level', measuring tapes and ranging poles described in Max Lock Group Nigeria reports (Theis and Coulthard, 1979, pp. 288–324). The process was used successfully for the site survey, layout design and setting out on the ground of layouts of 2,000 plots each in Biu, Buni Yadi and Geidam for their respective local governments and in Potiskum where two sites and services schemes for 4,000 plots including associated social and commercial facilities were implemented for the Fika local government. The plots and road reservations were all demarcated with permanent colour-coded steel tube markers at a sufficient accuracy for legal allocation purposes, that is with a half-metre certainty, very similar to the housing tolerances anywhere in the world. Apart from the design and management expertise of the town-planning team leaders, the staff used for this exercise were all untrained before some impromptu on-the-spot training. All these examples were then in Borno State and were carried out in 1977 and 1978.

The process worked well, was cost effective and most important of all was implemented without delay, enabling people to make their personal investments in housing securely and according to their circumstances. The 'intermediate layout' process became an example of how simplified implementation can respond to administrative demands; it also became a hallmark of the Max Lock Group approach. For example, the system was also used in setting out work at forty-seven village farm centre sites for the World Bank Agricultural Project in Bida, Niger State, Nigeria, in 1984.

Future housing: a practical strategy

In order to cope with future housing demand, the strategy should be to reap the benefit of both the social impetus for personal investment and the inherent cultural skills for building one's own home. It is the most feasible, cost-effective solution to the demand for housing. Given a framework of good urban planning, it also becomes the natural physical expression of the community and its social well-being.

To achieve this strategy, there needs to be a return to a recognised process for the designation and layout of sufficient suitable land for housing areas to meet the anticipated real demand. These two factors – designated land and real demand – are more complex than meets the eye.

The *'recognised process'* implies a framework of local adminstration, available to all, that would facilitate the allocation of housing land, administer any financial aid from the Housing Corporation or Federal Housing Authority and exercise the necessary safety controls. Such controls include appropriate building regulations and keeping a clear distinction between land for building and land to be kept for future infrastructure, floodland or farmland.

The quantification of *'sufficient suitable land'* implies a coordination of analysis of the urban area, in terms of physical constraints and opportunities for growth, with the design and availability of practical, cost-effective, basic infrastructure. This alone suggests a fairly sophisticated on-going interest in town planning and land management. It is necessary.

The quantification of *'anticipated real demand'* implies an understanding of the extent of numbers of households with a productive capacity for housebuilding. This in turn implies that a statistical data bank of such factors and their rates of change is held and up-dated regularly. This also suggests tasks that an urban administration should be undertaking as part of their town-planning operations. It, too, is vital.

The appropriateness of traditional housing and its success as a component of housing policy have their roots in affordability. To be part of everyday life once again, the management of land and revolving funds for basic infrastructure at a local level are vital.

Policy action guidelines

From the study of Katsina and elsewhere in northern Nigeria, the Max Lock Group approach has been to recommend a set of policy action guidelines (see Theis *et al.*, 1976; 1979; 1982; and Mutter, 1978; 1980). These guidelines are summarised here.

A policy of coordination is required between the various arms of government to reap the benefit of their individual contributions towards housing. The Federal Housing Authority could well undertake this coordinating role rather than invest in directly constructed housing themselves. Such built housing need only be stimulated by institutions such as universities or the police and other bodies that have some special circumstances. The existing government housing stock should be put on a non-subsidised rental basis, either in the private sector or as self-governing trusts at a local level where the tenants ensure the upkeep of maintenance; both would be overseen by local FHA staff.

The primary federal government role would be more effective in setting the national policies, priorities and guidelines within which the State Housing Corporations and local government can invest in the servicing of land for housing, both as improvement and upgrading schemes in existing built-up areas, and as sites-and-services schemes in expansion areas. The financial arrangements for government investment in such schemes would need to be balanced with the overall urban administration costs, to be borne by the town councils from their government grant and locally raised revenue from property holdings. The provision of urban services must be put on an accountable and on-going maintenance basis at the local level.

Investment funds saved by the federal government withdrawing from the current direct house-building programme should be used as grants to the Federal Mortgage Bank and state governments to promote loans to low-income groups and upgrading/sites-and-services schemes respectively. The Federal Mortgage Bank should revise its loan terms to make them suitable to a wider range of applicants. Similarly, commercial lending institutions should be encouraged to make housing loans available to low- and middle-income applicants. All private-sector developers should contribute to both the common and the particular infrastructure costs of housing projects.

Plot allocation procedures should be streamlined and made fully available. Where Planning Authorities do not exist, Corporate Planning and Implementation Committees should be established to administer upgrading. Layouts should simply allow for phased investment in infrastructure, starting at the minimum and progressing as finance and affordability permit on a revolving-fund basis. Community-based services should be provided initially and extended to each household as affordable. As a matter of principle the repayments from larger sites should bear the greater proportion; loan repayments should not need to exceed one-quarter of payees' income; and finance should be administered locally, on site, with advice on best building methods, materials, and controlled limits to development. Lastly, the use of local building materials should be encouraged by the adoption of a dual set of building regulations, one for manufactured or imported construction techniques and one for local traditional materials (Cox *et al.*, 1978).

Conclusion: a positive role for traditional housing in the process of rapid urbanisation

In the world of the 1990s, the global population is rising steadfastly – from an estimated 5.2 billion people in 1990 to approaching six billion by the year 2000 and to over eight billion projected by the year 2020, an increase of some 2.8 billion in thirty years. This growth itself is cause for concern, but its effects on cities in less-developed countries are multiplied by the rapid rise in levels of urbanisation. In 1990 the proportion of the global population living in towns was 43 per cent; in 2020, if current trends continue, with people being attracted to the greater productivity of urban areas, it will be 62 per cent. The greatest part of this increase will be in less-developed countries. By 2020 they will account for nearly 80 per cent of the world's urban population. This surge to the towns will need whatever assistance in organised development it can muster.

This dramatic increase of urbanisation is a natural part of the typical development process. It need not be a bad thing. Economically, people in towns are highly productive compared with their rural counterparts; this is part of the attraction of towns. This additional generation of urban wealth stimulates a growth in food production as a specialisation in rural areas. However, the rapid urbanisation process is not easy to manage effectively. It needs a straightforward and systematic administration by government for the ordered release of serviced land with security of tenure for people to make their personal investments in their housing and their livelihoods. Balanced layouts for housing, social and commercial developments are thus required as extensions to existing towns and as the basis of new settlements where appropriate. Given the land and the opportunity, people know best what they can afford and would like to build for their households. They know best how to determine their requirements within the social norms. They do need help in ensuring that their requirements conform with simple and basic public health and safety regulations.

Within this framework, the conclusion of this study is that people will build an evolution of their traditional house. It suits them best.

In Katsina, I believe that there is a healthy development process that will take the basic ingredients of the traditional Hausa house forward to the next century. It will evolve to reflect changing circumstances. Housing policy needs to continue to recognise the usefulness of traditional housing as a component in the future housing strategy of the city and its region. The governing administrative and professional inputs to this process must foster the art of the possible. Sensible practical standards need to be achieved. The resultant city will find its own expression in a built form that will reflect the aspirations and sincerity of the society it serves. I believe that it can thus continue to evolve as the city of character, intrigue and charm that I have had the good fortune to know so well since our 1977 study.

References

Blair, T.L., 1968, 'Kaduna! African city – western plan', *Architectural Association Quarterly*, **1**(1).

Cox, O. *et al.*, 1978, *Third World Urban Housing: Aspirations, Resources, Programmes, Projects*, Watford, Overseas Division, Building Research Establishment.

Isaacs, R.R. *et al.*, 1969, 'Kaduna: an international dialogue', *Architectural Association Quarterly*, **1**(4).

*Lock, C.M. and Theis, M.M., 1967, *Kaduna 1917/1967/2017: A Survey and Plan of the Capital Territory for the Government of Northern Nigeria*, London, Faber and Faber.

*Mutter, M.N. (ed.), 1978, *Katsina, Daura, Funtua, Malumfashi, Dutsin-Ma, Kankia and Mani Surveys and Development Plans for the Kaduna State Government*, Max Lock Group Nigeria/Kaduna State Government, Nigeria/Warminster Press.

*Mutter, M.N. (ed.), 1980, *Minna: Master Plan 1979–2000*, Nigeria, Max Lock Group Nigeria Ltd/Niger State Government.

Mutter, M.N., 1991, *A Study of the Work of the Max Lock Group 1942–1952*, MA Dissertation, London, University of Westminster.

O'Connor, A.M., 1983, *The African City*, London, Hutchinson.

Schwerdtfeger, F.W., 1982, *Traditional Housing in African Cities – a Comparative Study of Houses in Zaria, Ibadan, and Marrakech*, Chichester, John Wiley & Sons.

*Theis, M.M. *et al.*, 1976, *Maiduguri, Bauchi, Yola, Gombe, Potiskum, Nguru and Mubi: Surveys and Structure Plans for the North Eastern State Government* Max Lock Group Nigeria/North Eastern State Government, Nigeria, Warminster Press.

*Theis, M.M. and Coulthard, L.H. (eds), 1979, *Bama, Baga, Biu, Damaturu, Gashua, Geidam, Gwoza: Surveys and Development Plans for the Borno State Government*, Max Lock Group Nigeria/Borno State Government, Nigeria, Warminster Press.

*Theis, M.M. and Adenekan, S. (eds), 1982, *Bida & Kontagora: Master Plans for the Niger State Government*, Nigeria, Max Lock Group Nigeria Ltd/Niger State Government.

Urquhart, A.W., 1977, *Planned Urban Landscapes of Northern Nigeria: a Case Study of Zaria*, Zaria, Nigeria, Ahmadu Bello University Press.

*Nigerian State Government planning reports are available in the Max Lock Archive at the School of Urban Development and Planning, University of Westminster.

14 Conclusion

Stephen Wyn Williams

> A sort of black smoke covers the city. The sun seen through it is a disc without rays . . . The fetid, muddy waters, stained with a thousand colours by the factories they pass . . . wander slowly around the refuge of poverty. [De Tocqueville, quoted in Douglas, 1989]

Although one might assume that the above quotation is a vignette of the environmental problems facing a Third World city in the 1990s, it is in fact a description of Manchester in the 1830s. On the basis of such implicit parallels it would be tempting to suggest that there is some sort of inevitability or invariance concerning the relationships between urban growth and environmental deterioration. This, however, would be unwise. While there are broad conformities in the sequencing of events, such that urban growth creates physical changes in the urban environment which in turn impact on the human population, the broader (both national and international) political, economic and social structures and inequalities have a profound effect in determining the nature of specific outcomes in particular localities.

Such context-specific outcomes are well illustrated in the chapters of this volume which explore some of the central issues implicated in the relationship between environmental problems and the process of urbanisation in the Third World. Many of the contributions reflect a concern for issues with which geographers and others have been involved for some time. However, the volume can also be seen as contributing to a more recent body of literature which focuses on the growing environmental crisis in urban areas of the Third World. In this latter context it is perhaps surprising to note that despite the burgeoning literature on sustainable development, only a small proportion has focused on the environmental problems of towns and cities in developing countries. Furthermore, it is only recently that external support agencies have fully recognised the scale of the problem and the complexity of the challenges facing urban policy-makers and managers. Initiatives which are attempting to fill this lacuna can be found in collaborative programmes such as the UNDP/UNCHS/World Bank Urban Management Program (UMP), the UNDP/World Bank Metropolitan Environment Improvement Program (MEIP) and the UNCHS Sustainable Cities Programme (SCP). Furthermore, a major outcome of the

1992 UNCED Earth Summit was a broad-based agreement that pollution problems, environmental hazards and poverty will be priorities on the urban agenda as we approach the end of the millennium. One of the important outcomes of the Earth Summit was 'Agenda 21'. This was adopted by all governments attending the summit and provides guidelines concerning sustainable development in terms of policy and practice. A key recommendation was that cities should develop their own environmental strategies and action plans in the context of a 'local Agenda 21'. The need for such an approach is well stated by Bartone (1991, p. 411) who argues that 'a sustainable development approach is urgently needed in the management of the urban environment – an approach that ensures the contribution of Third World cities to national economies development while protecting the urban poor from the consequences of environmental deterioration. Environmental issues must be an integral part of all Third World urban policy'. On the basis of current population projections and without effective and appropriate interventions it is not difficult to envisage a situation whereby the environmental problems currently afflicting Third World cities will simultaneously intensify and become more extensive. Even with such interventions the task facing planners and policy-makers will be daunting. Currently, cities are absorbing two-thirds of the total population increase in developing countries, amounting to an additional 600 million urban dwellers over the next decade. By the year 2000 1.9 billion will live in the urban areas of the Third World; this figure could approach 3.5 billion by 2020. Furthermore, there will be a steady increase in the number of large cities within which environmental problems will be particularly acute. At the present time approximately half of the urban population of developing countries is located in 394 cities with a population over 500,000; by the year 2000 the number of such cities will have exceeded 500 (see United Nations, 1986). While recent data suggest that the growth rates of the largest cities are declining, this is not the case for secondary cities and towns which are already exhibiting a range of significant environmental problems.

The range of environmental problems covered in this volume have recently been referred to as the 'Brown Agenda' (see Cohen, 1993). As indicated in Figure 14.1 to a greater or lesser extent these are directly related to the development process: some problems may worsen or improve with income growth, while others may worsen then improve (World Bank, 1992). More specifically, as Table 14.1 demonstrates, the nature, range and mix of environmental problems found in any urban area is clearly influenced by the general level of development of the country concerned. This typology is important in that it highlights a concern expressed by a number of writers suggesting that pressure for environmental interventions from activists and other groups in Europe and North America tend to be heavily influenced by the environmental agenda set by

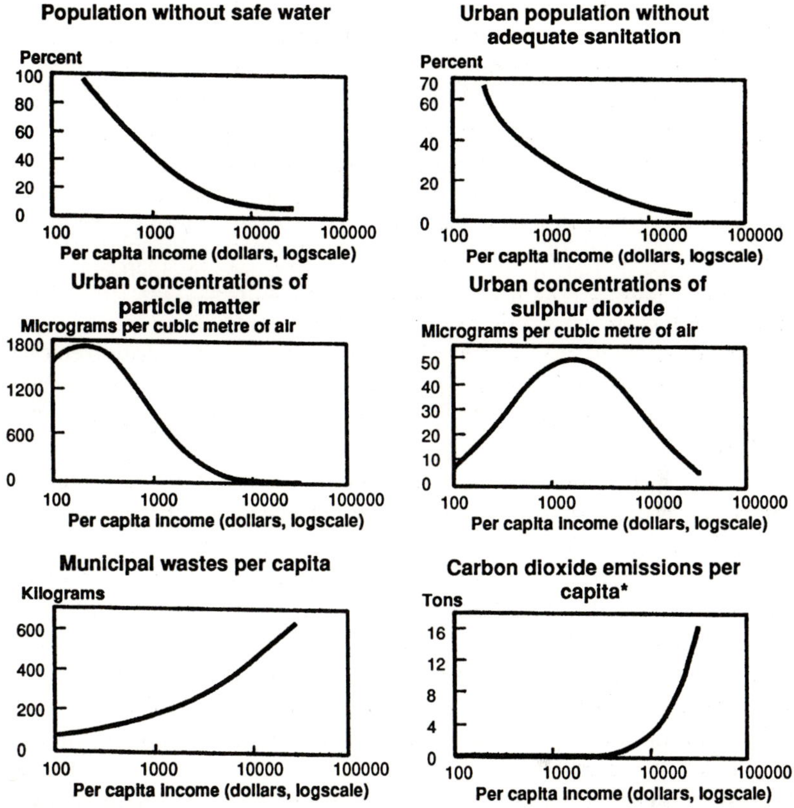

Note: Estimates are based on cross-country regression analysis of data from the 1980s.
* Emissions are from fossil fuels.

Figure 14.1 Environmental problems and the development process
Source: World Bank, 1992. Reproduced by permission

developed Northern cities. Consequently, as Hardoy *et al.* note, the environmental health agenda is frequently downplayed:

> . . . the particular problems which are documented may not be the most serious in terms of their impact on health and local ecosystems but simply those which are most easily documented or those which a particular research project or institution choose to document. One example is the fact that there is far more data on ambient city pollution levels for cities for (for instance) sulphur dioxide than of the extent to which city populations are served by sewers (or other effective sanitation systems) and storm drains. Yet the current health impact of inadequate sanitation and drainage is much greater than the health impact of current levels of sulphur dioxide. [1992, p. 25]

Table 14.1 Urban development and environmental problems

- *Low-income urban economies:* lack of basic environmental services such as water supply, sanitation, drainage, and solid-waste collection; a few temperate cities with severe air pollution from use of low-quality coal; uncontrolled land development and use; severe damage and loss of life from natural hazards.
- *Lower middle-income urban economies:* Low levels of sanitation and sewerage treatment; poor drainage; severe water pollution from municipal and industrial wastes; severe urban air pollution in some cities from vehicle emissions, use of low-quality fuels, and/or inefficient use of energy; inadequate solid-waste collection and disposal; lack of hazardous waste management; ineffective land-use controls; frequent disasters from natural and man-made hazards.
- *Upper-middle-income urban economies:* severe water pollution from municipal and industrial discharges; severe urban air pollution in some cities from vehicle emissions, use of low-quality fuels, and/or inefficient use of energy; inadequate solid-waste disposal, cumulative man-made hazards ('time bombs') – for example, hazardous waste dumps or nuclear power plants; frequent disasters resulting from natural and man-made hazards.
- *High-income urban economies:* health concerns from urban air pollution; loss of recreational amenities from water pollution; management of hazardous and toxic wastes; less subject to risks from natural and man-made hazards.

Source: Leitmann *et al.*, 1992

This does not imply that environmental problems associated with rapid industrialisation are unimportant but that we should recognise that the 'Brown Agenda' consists of two distinct components. The first of these is associated with traditional environmental health issues including the limited availability of land suitable for low-income groups and the lack of acceptable shelter and services to prevent the transmission of pathogens in the environment (see Misra, Chapter 11, in this volume). In Cairo, for example, while 70 per cent of the city is connected to the public sewerage system, only 15 per cent of the waste water collected is fully treated; 25 per cent is partially treated and 60 per cent is carried untreated for 200 kilometres by open canals to Lake Manzalah and then to the sea. The second component comprises those problems which have arisen owing to rapid industrialisation and at the present time tend to be less widespread than the first set of problems. These include toxic/hazardous wastes, water and air pollution, noise pollution and industrial accidents. However, for a growing number of cities in the Third World the overlap between these two sets of problems is becoming a critical issue (see Chan, Chapter 8, in this volume). That is, many cities are having to tackle more advanced pollution issues without having found effective solutions for 'traditional' pollution and health problems.

For those involved in formulating appropriate and realistic policies the important issues are understandably pragmatic in nature. Nevertheless, decisions at the intra-urban level have to address, and be made in the context of, a complex of underlying causes and their related effects. These include *inter alia*:

- Insufficient political will;
- A dynamic urbanisation process that outstrips the local capacity to provide needed environmental infrastructure and services;
- The underpricing of urban services resulting in their inefficient and inequitable distribution;
- Inefficient resource pricing (for example energy and water);
- Inadequate land-use control or inappropriate land-tenure systems coupled with excessive regulation of land markets;
- Inadequate institutional capacity for environmental management.

The issues noted above are certainly formidable; however as Rakodi and Devas (1993, p. 267) observe, most attempts to deal with the associated problems 'have been *ad hoc* responses to crisis situations such as grossly contaminated water supply or dangerous levels of air pollution'. The authors make the important point that in common with issues such as gender, it is not feasible to relegate environmental problems to sectoral concerns, but as yet they have not been integrated into planning and management structures for urban development. A second crucial issue in this context involves identifying interventions which correspond with the scale of the problem (see Table 14.2). This is by no means a straightforward task in that urban environmental problems often involve more than one environmental medium, a variety of economic sectors and a number of administrative jurisdictions. Even in the case of a clearly specified problem, such as contaminated water, interventions at a number of spatial scales would be required. Thus, at the household level a protected water supply would be required, which, in turn, would involve the creation of a water supply infrastructure at the community level. At the city level it would be necessary to provide resources and plans to enable action at power levels to be supported or undertaken. Finally, at the national level the government would need to ensure that the legislative framework was appropriate and that training opportunities were available (see Hardoy *et al.*, 1992, p. 152). However, as Douglass (1992) has illustrated, addressing these issues is a more complex problem than identifying appropriate levels for analysis and action. In a very real sense these levels of analysis are much more than administrative or organisational conveniences. Crucially, they represent social and political constituencies involving key questions concerning access to, and management of, economic and environmental resources. Consequently, environmental problems cannot simply be 'solved' by technocratic interventions but must involve much larger issues focusing on decentralisation, democratisation, empowerment and community-based involvement. As Hardoy and Satterthwaite put it: 'The achievement of an alternative

Table 14.2 Spatial scales and environmental problems

Spatial scale	Household workplace	Community	Metropolitan area	Region	Continent/ planet
Key infrastructure and Services	shelter; water storage; on site sanitation; garbage storage; stove ventilation	piped water; sewerage; garbage collection; drainage; streets/lanes	industrial parks; roads; interceptors; treatment plants; outfalls; landfills	highways; water sources; power plants	
Characteristic Problems	substandard housing; lack of water; no sanitation; disease vectors; indoor air pollution	excreta laden water/soils; trash dumping; flooding; noise/stress; natural disasters	traffic congestion; accidents; ambient air pollution; toxic dumps	water pollution; ecological areas lost	acid rain; global warming; ozone layer

Source: World Bank, 1993b

urban future depends on the extent to which poorer groups are able (or allowed) to organise not only within their district but also to become a greater political force within the city and the nation' (1990, p. 237). Douglass, underlines this view forcefully when he states 'that if low-income households and communities are not empowered to improve their capacity to manage the environment, the future of cities will be one of drastic environmental decline and immiseration that will also deeply affect the economies of cities and nations alike' (1992, p. 32).

The reciprocal and reinforcing relationships between housing, poverty and the environment has been a central theme in all the contributions to this volume. As Olpadwala and Goldsmith (1992, p. 630) note: 'The evidence today is overwhelming that the poor in cities suffer disproportionately from environmental problems.' Housing conditions are an unambiguous reflection of poverty, and in a recent World Bank policy paper the salient associations between housing conditions, poverty and environmental problems are clearly stated:

- The increased incidence of sickness and death is closely associated with housing that is too crowded, poorly built, located in unsafe areas or inadequately serviced with water and sanitation.
- Higher than expected prices combined with low incomes can result in high levels of overcrowding, low vacancy rates and homelessness.
- Poor quality housing is characterised by impermanence, fire-proneness, dilapidation and poor maintenance.
- Lack of residential infrastructure – intermittent or unsafe water supply, non-existent sewage, drainage and garbage disposal.
- Land tenure is insecure which can result in eviction and demolition.
- Housing is poorly located – either for economic opportunities or in unsafe locations subject to natural and man-made hazards.
- A close association exists between quality of the urban environment and the performance of the housing sector.
- Slums and squatter settlements are places of lower environmental quality. Polluted water, inadequate sanitation, indoor pollution caused by wood burning etc. are major causes of disease.
- Illegality reduces motivation to invest in improving quality of the local environment [World Bank 1993a, pp. 27–33]

Without doubt, the range of environmental problems associated with low-income groups requires interventions on a range of fronts. However, if one had to identify a single factor whose influence was paramount, then surely this must be the question of urban land and its management. As McAuslan (1985, p. 13) argues 'Land – its use, abuse, control and ownership – is the central problem of the city.' This is not the place to review the complexity of land-use policy in Third World cities. Nevertheless, some comments are necessary in so far as many of the issues so far discussed are related (directly or indirectly) to the land question. (For recent discussions on this theme see Baross and Van der Linden, 1990; Farvacque and McAuslan, 1992; Regional Development Dialogue, 1992; Mattingley, 1993.) Clearly, environmental quality is directly influenced by urban land-use decisions. Distortions in land markets and ineffective land-management policies and practices contribute substantially to the degradation of fragile lands, the occupation of hazardous locations, air pollution, congestion and the loss of open space and agricultural land. The corollary of this is the urgent need for effective urban land management which, in particular, would reduce the vulnerability of the urban poor to both natural and man-made hazards. In this context the central goal should be to improve land access for low-income groups.

In discussing this problem in Asian cities Sazanami *et al.* (1992) argue that two related issues need to be addressed. First, to secure land tenure for the poor who live on insecure land. Second, to increase land supply for affordable housing for low-income groups. (However, the tendency for land supplies to be increasingly commercialised in many cities frequently results in high- and middle-income groups competing for land made available for low-income groups. A parallel strategy of making available land for higher-income users would appear to be a necessary supply-side policy). In the first instance, a priority for governments would be to establish a system for clarifying land ownership and boundaries and to provide secure tenure in illegal settlements, including those in hazard-prone areas (for a discussion see McAuslan, 1992).

In terms of environment and housing this would enable landowners to gain credit and invest, for example, in improvements in dwellings so as to reduce vulnerability to floods, earthquakes and other natural hazards (see Degg, Chapter 2 in this volume). Secure tenure would also increase the motivation to demand improved infrastructure, thereby improving environmental health. The second issue (that is increasing the amount of land) is related to the broader question of reorganising the spatial patterns of land use in cities in an economically, socially and environmentally sustainable manner. As Bertraud (1992, p. 35) reminds us, 'a well-functioning land market is . . . essential in shaping the city and in the allocation of land among various urban activities in an efficient manner'. However, in many Third World cities a significant proportion of land is non-tradable, being set aside as a public good. According to the degree of regulatory control this can amount to between 20–70 per cent of land within municipal boundaries. This highly regulated environment can have severe impacts on the urban poor, whereby land development standards artificially reduce the supply and raise the price of land by requiring large lot sizes or land for transport and green spaces. However, there are dilemmas facing planners. For example:

> the imposition of properly enforced development restrictions on hazard-prone lands may substantially reduce the supply of developable land in a city and therefore increase the cost of available land for housing. Unless the government or private sector can provide alternative affordable housing sites, the increase in land prices will (i) push the poor to outlying areas where they will be even further away from employment opportunities and urban services or (ii) increase excessively the density of accessible inner city areas. By contrast, if the development restrictions are not enforced, the very delineation of hazard-prone areas will direct the poor right to them because these areas are often the only ones available for squatter settlement in accessible locations. [World Bank, 1993b]

As in many areas of activity in the Third World, deregulation would appear to provide the focus for an emerging consensus. Farvacque and McAuslan (1992, p. 105) propose that 'instead of the public sector having the dominant controlling role . . . it will move to a residual enabling role . . . the private sector will develop into the primary mode of land developing and development and will be facilitated to do this' (see also World Bank, 1993a). Clearly, for such a strategy to be economically and environmentally sustainable (other than at a purely local level) will require the development of an integrated policy approach. Furthermore, as Rakodi observes, a greater overt emphasis on the market will demand a more sophisticated analytical framework (1992, p. 43). This will involve a broad methodological and theoretical approach to land markets, focusing on prices and land supply, actors and institutions and the role of government and cultural factors in shaping property relations (see also the Fitzwilliam Memorandum, 1991).

Ultimately, any discussion which focuses on finding solutions to environmental problems in Third World cities cannot be divorced from the broader framework relating to the effective planning and management of urban areas. For some writers this involves developing a vision of an urban future which goes beyond immediate preoccupations. From an ecocentric perspective, Atkinson, for example, argues for a form of bioregionalism which 'is concerned . . . to fuse local culture with local environment in a context of modern, albeit appropriately redesigned, technology (1991, p. 411; see also Atkinson, 1992). The overwhelming problems facing urban areas, however, have meant that in practice a more pragmatic 'vision' concerned with achieving short- and medium-term goals is the reality. However, the achievement of even relatively modest goals can be decisively influenced by inappropriate or non-existent economic, legislative and institutional arrangements. Indeed, Hardoy *et al.* are of the opinion that it is the general failure of governance in Third World cities which is the root cause of many so-called environmental problems: '. . . remedying this failure . . . and addressing the reasons which underlie it should be central to any new urban environmental agenda' (1992, p. 205).

In the light of these rather pessimistic comments, the need to develop and implement appropriate policy responses is an even more urgent task.

However, the development of an approach to urban environmental management which is multisectoral, multidisciplinary and participatory will have to confront a number of fundamental issues. According to the World Bank (Leitman *et al.*, 1992) the most critical of these are the mobilisation of public support and participation, improving policy interventions and the operation of urban systems, building institutional capacity and closing the knowledge gap. In addressing these issues the ultimate goal would be to produce city-specific urban management programmes which would start with a participatory consultative phase. The second stage would consist of a long-term environmental strategy (EMS) which would guide a short- and medium-term environment action plan (EAP); the final outcome would be a sustainable investment programme (see Figure 14.2). Whether this represents another false start or a new dawn, only time will tell! What is clear, however, is that given the complexity of the issues involved, the resources available and the constraints which exist in most Third World cities any environmental management programme will need to be characterised by what Rakodi and Devas refer to as the 'New realism': '. . . this implies the need to adopt more appropriate and affordable standards, and to intervene more selectively. Over ambitious attempts to control not only lead to frustration and failure but also encourage corruption, as people seek ways around controls. The benefits, such as they are, of such misdirected intervention are likely to accrue to the privileged' (Rakodi and Devas, 1992 p. 270).

The approach suggested by the World Bank and other international agencies no doubt constitutes an emerging consensus concerning the environmental future of Third World cities. As such this represents a reaffirmation of technocentrism, albeit of the 'accommodating' variety, over ecocentrism. However, the possibility of alternative trajectories should not be dismissed. Referring to bioregionalism in particular, Atkinson suggests that we ignore these alternatives at our peril:

> Technocentric approaches to sustainability – and this is true of almost the entire 'sustainable cities' literature – take the view that we are not trying hard enough to implement poverty alleviation programmes and environmental controls; and that if we can put more resources into these, then all will be well. They hardly consider it necessary to generate criticism or create effective frameworks to judge whether particular policies are genuinely sustainable, and they are disinclined to participate in any enquiry into broader 'alternative development paradigms' or salient methodological – theoretical – philosophical questions . . . [1992, p. 328]
>
> From an 'environmental-realist' perspective that is convinced of the inviability in the medium term of the trajectory which humanity is presently taking, but which participates actively in the day to day world, there seems little hope that any effective rearrangement of social, political and cultural life will be affected before a series of increasingly severe environmental shocks. Some form of bioregionalism seems to be the inevitable outcome of this, if not arrived at by choice

Figure 14.2 The urban environmental planning process
Source: Leitmann *et al.*, 1992. Reproduced by permission

and design, then forced with great pain upon a reluctant world as the result of a catastrophic self-destruction of the Modern World System. [1992, p. 349]

Understandably, for planners and policy-makers grappling with the day-to-day problems of managing the unmanageable, the portentous views expressed in the above quotation represent a yawning chasm between the realities of the present and the uncertainties of the future. Nevertheless, our minds should not be closed to alternative paradigms, even if they might appear implausible at the present conjuncture. Indeed, if sustainable development is the goal, then the dominant growth paradigm which stresses consumerism and dominance over nature would seem to have little to offer as we approach the end of the millennium.

References

Atkinson, A., 1991, 'Environment and development: concepts and practices in transition', *Public Administration and Development*, **11**, pp. 401–13.

—, 1992, 'The urban bioregion as "Sustainable development" paradigm', *Third World Planning Review*, **14**(4), pp. 327–54.

Baross, P. and Van Der Linden, J. (eds), 1990, *The Transformation of Land Supply Systems in Third World cities*, Aldershot, Avebury.

Bartone, C., 1991, 'Environmental Challenges in Third World cities', *Journal of the American Planning Association*, **57**(4), pp. 411–415.

Bertaud, A., 1992, 'The impact of landuse regulations on land supply, consumption and price', *Regional Development Dialogue*, **13**(1), pp. 35–40.

Cohen, M., 1993, 'Megacities and the environment', *Finance and Development*, **30**(2), pp. 44–7.

Douglas, I., 1989, 'The rain on the roof: a geography of the urban environment' in D. Gregory and R. Walford (eds), *Horizons in Human Geography*, Basingstoke, Macmillan, pp. 217–38.

Douglass M., 1992, 'The political economy of urban poverty and environmental management in Asia: access, empowerment and community based alternatives', *Environment and Urbanisation*, **4**(2), pp. 9–32.

Farvacque, C. and McAuslan, P., 1992, *Reforming Urban Land Policies and Institutions in Developing Countries*, Urban Management Program Policy Paper 5, Washington, The World Bank.

Fitzwilliam Memorandum, 1991, 'International research workshop: land value changes and the impact of urban policy upon land valorisation processes in developing countries', *International Journal of Urban and Regional Research*, **15**(4), pp. 623–8.

Hardoy, J.E. and Satterthwaite, D., 1990, 'The future city' in J.E. Hardoy, S. Cairncross and D. Satterthwaite (eds), *The Poor Die Young*, London, Earthscan.

Hardoy, J.E., Mitlin, D. and Satterthwaite, D., 1992, *Environment Problems in Third World Cities*, London, Earthscan.

Leitman, J., Bartone, C. and Bernstein, J., 1992, 'Environmental management and urban development: issues and options for Third World cities', *Environment and Urbanisation*, **2**(4), pp. 131–40.

McAuslan, P., 1985, *Urban Land and Shelter for the Poor*, London, Earthscan.

—, 1992, 'Institutional/legal arrangements for the improved administration of land development, *Regional Development Dialogue*, **13**(1), pp. 14–32.

Mattingley, M., 1993, 'Urban management intervention in land markets' in N. Devas and C. Rakodi (eds), *Managing Fast Growing Cities*, Harlow, Longman, pp. 102–31.

Olpadwala, P. and Goldsmith, W.W., 1992, 'The sustainability of privilege: reflections on the environment, the Third World city, and poverty, *World Development*, **20**(4), pp. 627–40.

Rakodi, C., 1993, 'Planning for whom?' in N. Devas and C. Rakodi (eds), *Managing Fast Growing Cities*, Harlow, Longman, pp. 207–35.

Rakodi, C. and Devas, N. 1993, 'Conclusions: assessing the new approaches' in N. Devas and C. Rakodi (eds), *Managing Fast Growing Cities*, Harlow, Longman, pp. 265–96.

Regional Development dialogue, 1992, Theme issue: Urban Land Policies and Planning Systems, **13**(1).

Sazanami, H., Kidokoro, T. and Lim, Hooi Siang, 1992, 'Improving land access for the poor in Asian megacities', *Regional Development Dialogue*, **13**(1), pp. 3–13.

United Nations, 1986, *Urban and Rural Population Projections 1950–2025*, United Nations, New York

World Bank, 1992, *World Development Report*, New York, Oxford University Press.

—, 1993a, *Housing: Enabling Markets to Work*, A World Bank Policy Paper, Washington DC, The World Bank.

—, 1993b, *Toward Environmental Strategies for Cities*, Draft Strategy Framework Paper, Washington DC, The World Bank.

Place name index

Entries in *italics* refer to Figures; 'n' refers to endnotes

Subject index

Entries in *italic* refer to figures; 'n' refers to endnotes

The indexes were compiled by Dr Sabiha Sayeed